Inorganic Chemistry

Inorganic Chemistry

Edited by
Dennis Close

C WILLFORD PRESS

www.willfordpress.com

Published by Willford Press,
118-35 Queens Blvd., Suite 400,
Forest Hills, NY 11375, USA

ISBN: 978-1-68285-474-7

Cataloging-in-Publication Data

Inorganic chemistry / edited by Dennis Close.
    p. cm.
Includes bibliographical references and index.
ISBN 978-1-68285-474-7
1. Chemistry, Inorganic. 2. Inorganic compounds. 3. Chemistry.
I. Close, Dennis.
QD151.3 .I56 2018
546--dc23

For information on all Willford Press publications
visit our website at www.willfordpress.com

WILLFORD PRESS

Contents

Preface VII

Chapter 1   **Understanding Coordination Chemistry** 1

    i.   Coordination Complex 1

    ii.   Werner's Coordination Theory 13

    iii.   The Chelating Effect 15

    iv.   Valance Theory of Coordination Complexes 18

    v.   Crystal Field Theory 20

    vi.   Magnetochemistry 24

Chapter 2   **Bioinorganic Chemistry: An Overview** 35

    i.   Bioinorganic Chemistry 35

    ii.   Biological Essential Elements with their Functions 39

    iii.   Na+/K+-ATPase 43

    iv.   Photosystems 49

    v.   Cytochrome c Oxidase 52

    vi.   Iron-sulfur Proteins and Nitrogenase 55

    vii.   Hemocyanine 58

    viii.   Alcohol Dehydrogenase 61

Chapter 3   **Chemical Bonding in Inorganic Chemistry** 64

    i.   Chemical Bonding 64

    ii.   Bonding 71

    iii.   Lattice Energy 89

    iv.   Covalent Bond 96

    v.   VSEPR Theory 103

    vi.   Molecular Orbital Theory 114

Chapter 4   **Acids-Bases and its Related Theories** 122

    i.   Acid–base Reaction 122

    ii.   The Arrhenius Theory 130

    iii.   Franklin Theory (1905) 130

    iv.   Brønsted–Lowry Acid–base Theory 131

    v.   Lewis Acids and Bases 136

    vi.   HSAB Theory 143

    vii.   Strength of Lewis Acids and Bases 148

Chapter 5   **Reduction–Oxidation Reactions** 153

    i.   Redox 153

    ii.   Latimer Diagram 173

    iii.   Frost Diagram 177

    iv.   Nernst Equation 182

    v.   Redox Dependence 189

Permissions

Index

Preface

Inorganic chemistry refers to that branch of chemistry, which deals with the study and analysis of all inorganic and organometallic compounds. It can be further divided into inorganic chemistry and theoretical inorganic chemistry. Inorganic chemistry finds its application in fields like agriculture, materials science, surfactants, catalysis, medications, fuel, etc. This book elucidates the concepts and innovative models around prospective developments with respect to inorganic chemistry. It presents this complex subject in the most comprehensible and easy to understand language. The topics covered herein offer the readers new insights in this field. This textbook will serve as a valuable source of reference for those interested in this subject.

To facilitate a deeper understanding of the contents of this book a short introduction of every chapter is written below:

Chapter 1- A coordination complex consists of a metallic ion or atom and a surrounding array of ligands. The central atom or coordinating center may or may not be metallic, and is called a metal complex when it is. The study of such compounds is called coordination chemistry. This chapter is an overview of the subject matter incorporating all the major aspects of coordination chemistry.

Chapter 2- Bioinorganic chemistry analyses the role of metals in biology. It can be said to be an interdisciplinary subject composing of biochemistry and inorganic chemistry. It studies biological processes such as ion transport, mutation, etc. The diverse applications of bioinorganic chemistry in the current scenario have been thoroughly discussed in this chapter.

Chapter 3- Chemical bonds are the forces that bind atoms together to form a chemical compound. Chemical bonding is caused through electrostatic force or through the sharing of electrons as seen in covalent bonds. The chapter closely examines the key concepts of chemical bonding to provide an extensive understanding of the subject.

Chapter 4- A chemical reaction occurring between a base and an acid is known as an acid-base reaction. Theories that provide conceptual frameworks to the reaction mechanisms of acid-base reactions include the Arrhenius theory, the Brønsted–Lowry acid–base theory and the Franklin theory. The topics discussed in the chapter are of great importance to broaden the existing knowledge on inorganic chemistry.

Chapter 5- A reaction where the oxidation states of atoms are changed are known as reduction-oxidation reactions or redox reactions. Redox potential is the discerning of the driving force of the reaction as being either reductive or a change in states. The chapter on reduction-oxidation reactions offers an insightful focus, keeping in mind the complex subject matter.

I would like to share the credit of this book with my editorial team who worked tirelessly on this book. I owe the completion of this book to the never-ending support of my family, who supported me throughout the project.

Editor

Understanding Coordination Chemistry

A coordination complex consists of a metallic ion or atom and a surrounding array of ligands. The central atom or coordinating center may or may not be metallic, and is called a metal complex when it is. The study of such compounds is called coordination chemistry. This chapter is an overview of the subject matter incorporating all the major aspects of coordination chemistry.

Coordination Complex

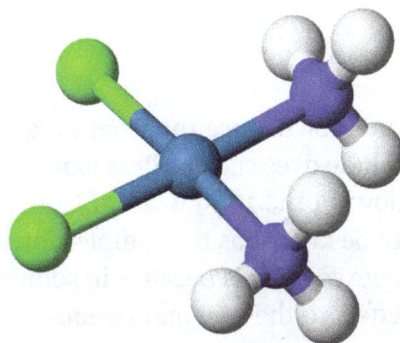

Cisplatin, $PtCl_2(NH_3)_2$, is a coordination complex of platinum(II) with two chloride and two ammonia ligands. It is one of the most successful anticancer drugs.

In chemistry, a coordination complex consists of a central atom or ion, which is usually metallic and is called the *coordination centre*, and a surrounding array of bound molecules or ions, that are in turn known as *ligands* or complexing agents. Many metal-containing compounds, especially those of transition metals, are coordination complexes. A coordination complex whose centre is a metal atom is called a metal complex.

Nomenclature and Terminology

Coordination complexes are so pervasive that their structures and reactions are described in many ways, sometimes confusingly. The atom within a ligand that is bonded to the central metal atom or ion is called the donor atom. In a typical complex, a metal ion is bonded to several donor atoms, which can be the same or different. A polydentate (multiple bonded) ligand is a molecule or ion that bonds to the central atom through several of the ligand's atoms; ligands with 2, 3, 4 or even 6 bonds to the central atom are common. These complexes are called chelate complexes, the formation of such complexes is called chelation, complexation, and coordination.

The central atom or ion, together with all ligands comprise the coordination sphere. The central atoms or ion and the donor atoms comprise the first coordination sphere.

Coordination refers to the "coordinate covalent bonds" (dipolar bonds) between the ligands and the central atom. Originally, a complex implied a reversible association of molecules, atoms, or ions through such weak chemical bonds. As applied to coordination chemistry, this meaning has evolved. Some metal complexes are formed virtually irreversibly and many are bound together by bonds that are quite strong.

The number of donor atoms attached to the central atom or ion is called the coordination number. The most common coordination numbers are 2, 4 and especially 6. A hydrated ion is one kind of a complex ion (or simply a complex), a species formed between a central metal ion and one or more surrounding ligands, molecules or ions that contain at least one lone pair of electrons,

If all the ligands are monodentate, then the number of donor atoms equals the number of ligands. For example, the cobalt(II) hexahydrate ion or the hexaaquacobalt(II) ion $[Co(II_2O)_6]^{2+}$, is a hydrated-complex ion that consists of six water molecules attached to a metal ion Co. The oxidation state and the coordination number reflect the number of bonds formed between the metal ion and the ligands in the complex ion. However the coordination number of $Pt(en)_2^{2+}$ is 4 (rather than 2) since it has two bidentate ligands, which contain four donor atoms in total.

History

Coordination complexes have been known since the beginning of modern chemistry. Early well-known coordination complexes include dyes such as Prussian blue. Their properties were first well understood in the late 1800s, following the 1869 work of Christian Wilhelm Blomstrand. Blomstrand developed what has come to be known as the complex ion chain theory. The theory claimed that the reason coordination complexes form is because in solution, ions would be bound via ammonia chains. He compared this effect to the way that various carbohydrate chains form.

Alfred Werner

Following this theory, Danish scientist Sophus Mads Jørgensen made improvements to it. In his version of the theory, Jorgensen claimed that when a molecule dissociates in a solution there were two possible outcomes: the ions would bind via the ammonia chains Blomstrand had described or the ions would bind directly to the metal.

It was not until 1893 that the most widely accepted version of the theory today was published by Alfred Werner. Werner's work included two important changes to the Blomstrand theory. The first was that Werner described the two different ion possibilities in terms of location in the coordination sphere. He claimed that if the ions were to form a chain this would occur outside of the coordination sphere while the ions that bound directly to the metal would do so within the coordination sphere. In one of Werner's most important discoveries however he disproved the majority of the chain theory. Werner was able to discover the spatial arrangements of the ligands that were involved in the formation of the complex hexacoordinate cobalt. His theory allows one to understand the difference between a coordinated ligand and a charge balancing ion in a compound, for example the chloride ion in the cobaltammine chlorides and to explain many of the previously inexplicable isomers.

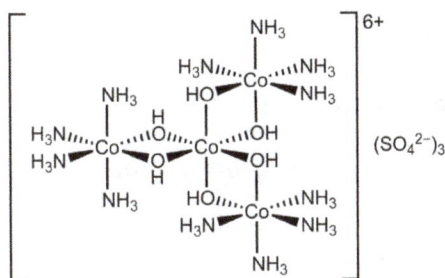

Structure of hexol

In 1914, Werner first resolved the coordination complex, called hexol, into optical isomers, overthrowing the theory that only carbon compounds could possess chirality.

Structures

The ions or molecules surrounding the central atom are called ligands. Ligands are generally bound to the central atom by a coordinate covalent bond (donating electrons from a lone electron pair into an empty metal orbital), and are said to be coordinated to the atom. There are also organic ligands such as alkenes whose pi bonds can coordinate to empty metal orbitals. An example is ethene in the complex known as Zeise's salt, $K^+[PtCl_3(C_2H_4)]^-$.

Geometry

In coordination chemistry, a structure is first described by its coordination number, the number of ligands attached to the metal (more specifically, the number of donor atoms). Usually one can count the ligands attached, but sometimes even the counting can become ambiguous. Coordination numbers are normally between two and nine, but large numbers of ligands are not uncommon for the lanthanides and actinides. The number of bonds depends on the size, charge, and electron configuration of the metal ion and the ligands. Metal ions may have more than one coordination number.

Typically the chemistry of transition metal complexes is dominated by interactions between s and p molecular orbitals of the ligands and the d orbitals of the metal ions. The s, p, and d orbitals of the metal can accommodate 18 electrons. The maximum coordination number for a certain metal is thus related to the electronic configuration of the metal ion (to be more specific, the number of empty orbitals) and to the ratio of the size of the ligands and the metal ion. Large metals and small ligands lead to high coordination numbers, e.g. $[Mo(CN)_8]^{4-}$. Small metals with large ligands lead to low coordination numbers, e.g. Pt[P(CMe$_3$)]$_2$. Due to their large size, lanthanides, actinides, and early transition metals tend to have high coordination numbers.

Different ligand structural arrangements result from the coordination number. Most structures follow the points-on-a-sphere pattern (or, as if the central atom were in the middle of a polyhedron where the corners of that shape are the locations of the ligands), where orbital overlap (between ligand and metal orbitals) and ligand-ligand repulsions tend to lead to certain regular geometries. The most observed geometries are listed below. There are cases that deviate from a regular geometry due to the use of ligands of different types (which results in irregular bond lengths) or due to the size of ligands.

- Linear for two-coordination

- Trigonal planar for three-coordination

- Tetrahedral or square planar for four-coordination

- Trigonal bipyramidal or square pyramidal for five-coordination

- Octahedral (orthogonal) for six-coordination

- Pentagonal bipyramidal, capped octahedral or capped trigonal prismatic for seven-coordination

- Square antiprismatic or dodecahedral for eight-coordination

- Tri-capped trigonal prismatic (triaugmented triangular prism) or capped square antiprismatic for nine-coordination.

Due to special electronic effects such as (second-order) Jahn–Teller stabilization, certain geometries (in which the coordination atoms do not follow a points-on-a-sphere pattern) are stabilized relative to the other possibilities, e.g. for some compounds the trigonal prismatic geometry is stabilized relative to octahedral structures for six-coordination.

- Bent for two-coordination

- Trigonal pyramidal for three-coordination

- Trigonal prismatic for six-coordination

Isomerism

The arrangement of the ligands is fixed for a given complex, but in some cases it is mutable by a reaction that forms another stable isomer.

There exist many kinds of isomerism in coordination complexes, just as in many other compounds.

Stereoisomerism

Stereoisomerism occurs with the same bonds in different orientations relative to one another. Stereoisomerism can be further classified into:

Cis–trans Isomerism and Facial–meridional Isomerism

Cis–trans isomerism occurs in octahedral and square planar complexes (but not tetrahedral). When two ligands are adjacent they are said to be cis, when opposite each other, trans. When three identical ligands occupy one face of an octahedron, the isomer is said to be facial, or fac. In a fac isomer, any two identical ligands are adjacent or cis to each other. If these three ligands and the metal ion are in one plane, the isomer is said to be meridional, or mer. A *mer* isomer can be considered as a combination of a *trans* and a *cis*, since it contains both trans and cis pairs of identical ligands.

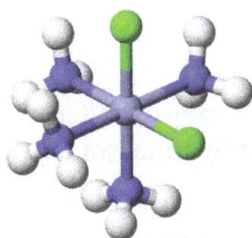
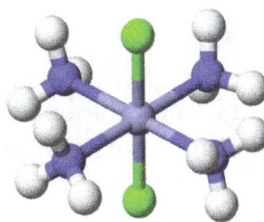

cis-[CoCl$_2$(NH$_3$)$_4$]$^+$ *trans*-[CoCl$_2$(NH$_3$)$_4$]$^+$

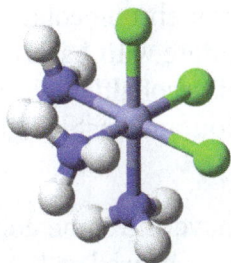
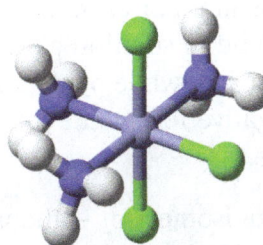

fac-[CoCl$_3$(NH$_3$)$_3$]   *mer*-[CoCl$_3$(NH$_3$)$_3$]

Optical Isomerism

Optical isomerism occurs when a molecule is not superimposable with its mirror image. It is so called because the two isomers are each optically active, that is, they rotate the plane of polarized light in opposite directions. The symbol Λ (*lambda*) is used as a prefix to describe the left-handed propeller twist formed by three bidentate ligands, as shown. Likewise, the symbol Δ (*delta*) is used as a prefix for the right-handed propeller twist.

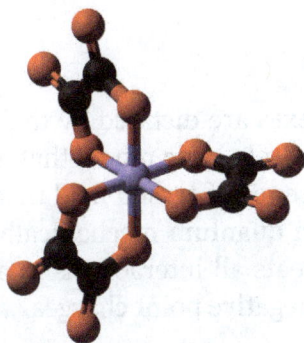
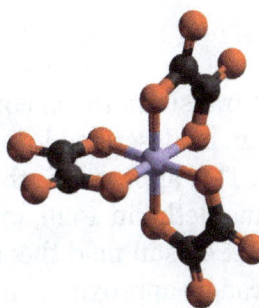

Λ-[Fe(ox)$_3$]$^{3-}$ Δ-[Fe(ox)$_3$]$^{3-}$

Λ-*cis*-[CoCl$_2$(en)$_2$]$^+$ $\qquad\qquad$ Δ-*cis*-[CoCl$_2$(en)$_2$]$^+$

Structural Isomerism

Structural isomerism occurs when the bonds are themselves different. There are four types of structural isomerism: ionisation isomerism, solvate or hydrate isomerism, linkage isomerism and coordination isomerism.

1. Ionisation isomerism – the isomers give different ions in solution although they have the same composition. This type of isomerism occurs when the counter ion of the complex is also a potential ligand. For example, pentaamminebromocobalt(III) sulfate [Co(NH$_3$)$_5$Br] SO$_4$ is red violet and in solution gives a precipitate with barium chloride, confirming the presence of sulfate ion, while pentaamminesulfatecobalt(III) bromide [Co(NH$_3$)$_5$SO$_4$]Br is red and tests negative for sulfate ion in solution, but instead gives a precipitate of AgBr with silver nitrate.

2. Solvate or hydrate isomerism – the isomers have the same composition but differ with respect to the number of solvent ligand molecules as well as the counter ion in the crystal lattice. For example, [Cr(H$_2$O)$_6$]Cl$_3$ is violet colored, [CrCl(H$_2$O)$_5$]Cl$_2$·H$_2$O is blue-green, and [CrCl$_2$(H$_2$O)$_4$]Cl·2H$_2$O is dark green.

3. Linkage isomerism occurs with ambidentate ligands that can bind in more than one place. For example, NO$_2$ is an ambidentate ligand: It can bind to a metal at either the N atom or an O atom.

4. Coordination isomerism – this occurs when both positive and negative ions of a salt are complex ions and the two isomers differ in the distribution of ligands between the cation and the anion. For example, [Co(NH$_3$)$_6$][Cr(CN)$_6$] and [Cr(NH$_3$)$_6$][Co(CN)$_6$].

Electronic Properties

Many of the properties of transition metal complexes are dictated by their electronic structures. The electronic structure can be described by a relatively ionic model that ascribes formal charges to the metals and ligands. This approach is the essence of crystal field theory (CFT). Crystal field theory, introduced by Hans Bethe in 1929, gives a quantum mechanically based attempt at understanding complexes. But crystal field theory treats all interactions in a complex as ionic and assumes that the ligands can be approximated by negative point charges.

More sophisticated models embrace covalency, and this approach is described by ligand field theory (LFT) and Molecular orbital theory (MO). Ligand field theory, introduced in 1935 and built from molecular orbital theory, can handle a broader range of complexes and can explain complexes in which the interactions are covalent. The chemical applications of group theory can aid in the understanding of crystal or ligand field theory, by allowing simple, symmetry based solutions to the formal equations.

Chemists tend to employ the simplest model required to predict the properties of interest; for this reason, CFT has been a favorite for the discussions when possible. MO and LF theories are more complicated, but provide a more realistic perspective.

The electronic configuration of the complexes gives them some important properties:

Color of Transition Metal Complexes

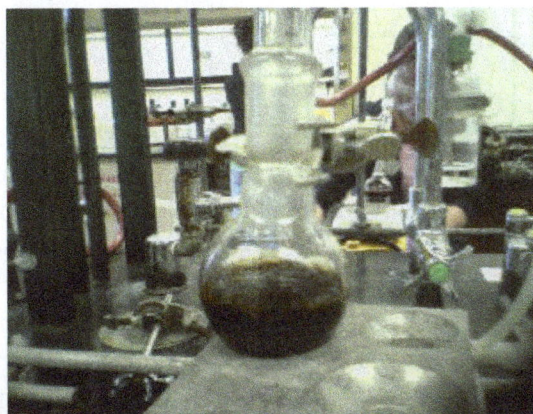

Synthesis of copper(II)-tetraphenylporphyrin, a metal complex, from tetraphenylporphyrin and copper(II) acetate monohydrate.

Transition metal complexes often have spectacular colors caused by electronic transitions by the absorption of light. For this reason they are often applied as pigments. Most transitions that are related to colored metal complexes are either d–d transitions or charge transfer bands. In a d–d transition, an electron in a d orbital on the metal is excited by a photon to another d orbital of higher energy. A charge transfer band entails promotion of an electron from a metal-based orbital into an empty ligand-based orbital (Metal-to-Ligand Charge Transfer or MLCT). The converse also occurs: excitation of an electron in a ligand-based orbital into an empty metal-based orbital (Ligand to Metal Charge Transfer or LMCT). These phenomena can be observed with the aid of electronic spectroscopy; also known as UV-Vis. For simple compounds with high symmetry, the d–d transitions can be assigned using Tanabe–Sugano diagrams. These assignments are gaining increased support with computational chemistry.

Colours of Various Example Coordination Complexes						
	Fe^{2+}	Fe^{3+}	Co^{2+}	Cu^{2+}	Al^{3+}	Cr^{3+}
Hydrated Ion	$[Fe(H_2O)_6]^{2+}$ Pale green Solution	$[Fe(H_2O)_6]^{3+}$ Yellow/brown Solution	$[Co(H_2O)_6]^{2+}$ Pink Solution	$[Cu(H_2O)_6]^{2+}$ Blue Solution	$[Al(H_2O)_6]^{3+}$ Colourless Solution	$[Cr(H_2O)_6]^{3+}$ Green Solution
OH⁻, dilute	$[Fe(H_2O)_4(OH)_2]$ Dark green Precipitate	$[Fe(H_2O)_3(OH)_3]$ Brown Precipitate	$[Co(H_2O)_4(OH)_2]$ Blue/green Precipitate	$[Cu(H_2O)_4(OH)_2]$ Blue Precipitate	$[Al(H_2O)_3(OH)_3]$ White Precipitate	$[Cr(H_2O)_3(OH)_3]$ Green Precipitate

OH⁻, concentrated	$[Fe(H_2O)_4(OH)_2]$ Dark green Precipitate	$[Fe(H_2O)_3(OH)_3]$ Brown Precipitate	$[Co(H_2O)_4(OH)_2]$ Blue/green Precipitate	$[Cu(H_2O)_4(OH)_2]$ Blue Precipitate	$[Al(OH)_4]^-$ Colourless Solution	$[Cr(OH)_6]^{3-}$ Green Solution
NH₃, dilute	$[Fe(H_2O)_4(OH)_2]$ Dark green Precipitate	$[Fe(H_2O)_3(OH)_3]$ Brown Precipitate	$[Co(H_2O)_4(OH)_2]$ Blue/green Precipitate	$[Cu(H_2O)_4(OH)_2]$ Blue Precipitate	$[Al(H_2O)_3(OH)_3]$ White Precipitate	$[Cr(H_2O)_3(OH)_3]$ Green Precipitate
NH₃, concentrated	$[Fe(H_2O)_4(OH)_2]$ Dark green Precipitate	$[Fe(H_2O)_3(OH)_3]$ Brown Precipitate	$[Co(NH_3)_6]^{2+}$ Straw coloured Solution	$[Cu(NH_3)_4(H_2O)_2]^{2+}$ Deep blue Solution	$[Al(H_2O)_3(OH)_3]$ White Precipitate	$[Cr(NH_3)_6]^{3+}$ Purple Solution
CO₃²⁻	$FeCO_3$ Dark green Precipitate	$[Fe(H_2O)_3(OH)_3]$ Brown Precipitate + bubbles	$CoCO_3$ Pink Precipitate	$CuCO_3$ Blue/green Precipitate		

Colors of Lanthanide Complexes

Superficially lanthanide complexes are similar to those of the transition metals in that some are coloured. However, for the common Ln³⁺ ions (Ln = lanthanide) the colors are all pale, and hardly influenced by the nature of the ligand. The colors are due to 4f electron transitions. As the 4f orbitals in lanthanides are "buried" in the xenon core and shielded from the ligand by the 5s and 5p orbitals they are therefore not influenced by the ligands to any great extent leading to a much smaller crystal field splitting than in the transition metals. The absorption spectra of an Ln³⁺ ion approximates to that of the free ion where the electronic states are described by spin-orbit coupling (also called L-S coupling or Russell-Saunders coupling). This contrasts to the transition metals where the ground state is split by the crystal field. Absorptions for Ln³⁺ are weak as electric dipole transitions are parity forbidden (Laporte Rule forbidden) but can gain intensity due to the effect of a low-symmetry ligand field or mixing with higher electronic states (*e.g.* d orbitals). Also absorption bands are extremely sharp which contrasts with those observed for transition metals which generally have broad bands. This can lead to extremely unusual effects, such as significant color changes under different forms of lighting.

Magnetism

Metal complexes that have unpaired electrons are magnetic. Considering only monometallic complexes, unpaired electrons arise because the complex has an odd number of electrons or because electron pairing is destabilized. Thus, monomeric Ti(III) species have one "d-electron" and must be (para)magnetic, regardless of the geometry or the nature of the ligands. Ti(II), with two d-electrons, forms some complexes that have two unpaired electrons and others with none. This effect is illustrated by the compounds $TiX_2[(CH_3)_2PCH_2CH_2P(CH_3)_2]_2$: when X = Cl, the complex is paramagnetic (high-spin configuration), whereas when X = CH_3, it is diamagnetic (low-spin configuration). It is important to realize that ligands provide an important means of adjusting the ground state properties.

In bi- and polymetallic complexes, in which the individual centres have an odd number of electrons or that are high-spin, the situation is more complicated. If there is interaction (either direct or through ligand) between the two (or more) metal centres, the electrons may couple (antiferromagnetic coupling, resulting in a diamagnetic compound), or they may enhance each other (ferromagnetic coupling). When there is no interaction, the two (or more) individual metal centers behave as if in two separate molecules.

Reactivity

Complexes show a variety of possible reactivities:

- Electron transfers

  A common reaction between coordination complexes involving ligands are inner and outer sphere electron transfers. They are two different mechanisms of electron transfer redox reactions, largely defined by the late Henry Taube. In an inner sphere reaction, a ligand with two lone electron pairs acts as a *bridging ligand*, a ligand to which both coordination centres can bond. Through this, electrons are transferred from one centre to another.

- (Degenerate) ligand exchange

  One important indicator of reactivity is the rate of degenerate exchange of ligands. For example, the rate of interchange of coordinate water in $[M(H_2O)_6]^{n+}$ complexes varies over 20 orders of magnitude. Complexes where the ligands are released and rebound rapidly are classified as labile. Such labile complexes can be quite stable thermodynamically. Typical labile metal complexes either have low-charge (Na^+), electrons in d-orbitals that are antibonding with respect to the ligands (Zn^{2+}), or lack covalency (Ln^{3+}, where Ln is any lanthanide). The lability of a metal complex also depends on the high-spin vs. low-spin configurations when such is possible. Thus, high-spin Fe(II) and Co(III) form labile complexes, whereas low-spin analogues are inert. Cr(III) can exist only in the low-spin state (quartet), which is inert because of its high formal oxidation state, absence of electrons in orbitals that are M−L antibonding, plus some "ligand field stabilization" associated with the d^3 configuration.

- Associative processes

  Complexes that have unfilled or half-filled orbitals often show the capability to react with substrates. Most substrates have a singlet ground-state; that is, they have lone electron pairs (e.g., water, amines, ethers), so these substrates need an empty orbital to be able to react with a metal centre. Some substrates (e.g., molecular oxygen) have a triplet ground state, which results that metals with half-filled orbitals have a tendency to react with such substrates (it must be said that the dioxygen molecule also has lone pairs, so it is also capable to react as a 'normal' Lewis base).

If the ligands around the metal are carefully chosen, the metal can aid in (stoichiometric or catalytic) transformations of molecules or be used as a sensor.

Classification

Metal complexes, also known as coordination compounds, include all metal compounds, aside from metal vapors, plasmas, and alloys. The study of "coordination chemistry" is the study of "inorganic chemistry" of all alkali and alkaline earth metals, transition metals, lanthanides, actinides, and metalloids. Thus, coordination chemistry is the chemistry of the majority of the periodic table. Metals and metal ions exist, in the condensed phases at least, only surrounded by ligands.

The areas of coordination chemistry can be classified according to the nature of the ligands, in broad terms:

- Classical (or "Werner Complexes"): Ligands in classical coordination chemistry bind to metals, almost exclusively, via their "lone pairs" of electrons residing on the main group atoms of the ligand. Typical ligands are H_2O, NH_3, Cl^-, CN^-, en. Some of the simplest members of such complexes are described in metal aquo complexes, metal ammine complexes.

  Examples: $[Co(EDTA)]^-$, $[Co(NH_3)_6]Cl_3$, $[Fe(C_2O_4)_3]K_3$

- Organometallic Chemistry: Ligands are organic (alkenes, alkynes, alkyls) as well as "organic-like" ligands such as phosphines, hydride, and CO.

  Example: $(C_5H_5)Fe(CO)_2CH_3$

- Bioinorganic Chemistry: Ligands are those provided by nature, especially including the side chains of amino acids, and many cofactors such as porphyrins.

  Example: hemoglobin contains heme, a porphyrin complex of iron

  Example: chlorophyll contains a porphyrin complex of magnesium

  Many natural ligands are "classical" especially including water.

- Cluster Chemistry: Ligands are all of the above also include other metals as ligands.

  Example $Ru_3(CO)_{12}$

- In some cases there are combinations of different fields:

  Example: $[Fe_4S_4(Scysteinyl)_4]^{2-}$, in which a cluster is embedded in a biologically active species.

Mineralogy, materials science, and solid state chemistry – as they apply to metal ions – are subsets of coordination chemistry in the sense that the metals are surrounded by ligands. In many cases these ligands are oxides or sulfides, but the metals are coordinated nonetheless, and the principles and guidelines discussed below apply. In hydrates, at least some of the ligands are water molecules. It is true that the focus of mineralogy, materials science, and solid state chemistry differs from the usual focus of coordination or inorganic chemistry. The former are concerned primarily with polymeric structures, properties arising from a collective effects of many highly interconnected metals. In contrast, coordination chemistry focuses on reactivity and properties of complexes containing individual metal atoms or small ensembles of metal atoms.

Naming Complexes

The basic procedure for naming a complex:

1. When naming a complex ion, the ligands are named before the metal ion.

2. Write the names of the ligands in alphabetical order. (Numerical prefixes do not affect the order.)

   o Multiple occurring monodentate ligands receive a prefix according to the number of occurrences: *di-*, *tri-*, *tetra-*, *penta-*, or *hexa*. Polydentate ligands (e.g., ethylenediamine, oxalate) receive *bis-*, *tris-*, *tetrakis-*, etc.

- o Anions end in *o*. This replaces the final 'e' when the anion ends with '-ide', '-ate' or '-ite', e.g. *chloride* becomes *chlorido* and *sulfate* becomes *sulfato*. Formerly, '-ide' was changed to '-o' (e.g. *chloro* and *cyano*), but this rule has been modified in the 2005 IUPAC recommendations and the correct forms for these ligands are now *chlorido* and *cyanido*.

- o Neutral ligands are given their usual name, with some exceptions: NH_3 becomes *ammine*; H_2O becomes *aqua* or *aquo*; CO becomes *carbonyl*; NO becomes *nitrosyl*.

3. Write the name of the central atom/ion. If the complex is an anion, the central atom's name will end in *-ate*, and its Latin name will be used if available (except for mercury).

1. If the central atom's oxidation state needs to be specified (when it is one of several possible, or zero), write it as a Roman numeral (or 0) in parentheses.

2. Name cation then anion as separate words (if applicable, as in last example)

Examples:

metal	changed to
cobalt	cobaltate
aluminium	aluminate
chromium	chromate
vanadium	vanadate
copper	cuprate
iron	ferrate

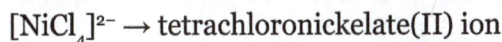

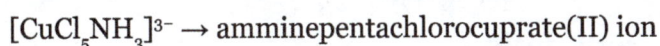

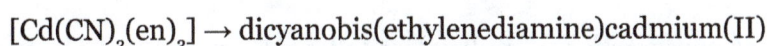

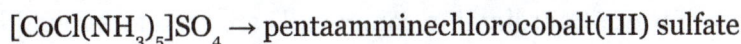

$[NiCl_4]^{2-} \rightarrow$ tetrachloronickelate(II) ion

$[CuCl_5NH_3]^{3-} \rightarrow$ amminepentachlorocuprate(II) ion

$[Cd(CN)_2(en)_2] \rightarrow$ dicyanobis(ethylenediamine)cadmium(II)

$[CoCl(NH_3)_5]SO_4 \rightarrow$ pentaamminechlorocobalt(III) sulfate

The coordination number of ligands attached to more than one metal (bridging ligands) is indicated by a subscript to the Greek symbol μ placed before the ligand name. Thus the dimer of aluminium trichloride is described by $Al_2Cl_4(\mu_2\text{-}Cl)_2$.

Stability Constant

The affinity of metal ions for ligands is described by stability constant. This constant, also referred to as the formation constant, is given the notation of K_f and can be calculated through the following method for simple cases:

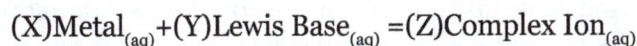

$(X)Metal_{(aq)}+(Y)Lewis\ Base_{(aq)} =(Z)Complex\ Ion_{(aq)}$

$$K_f = \frac{[Complex\ ion]^Z}{[Metal\ ion]^X[Lewis\ base]^Y}$$

Formation constants vary widely. Large values indicate that the metal has high affinity for the ligand, provided the system is at equilibrium.

Sometimes the stability constant will be in a different form known as the constant of destability. This constant is expressed as the inverse of the constant of formation and is denoted as $K_d = 1/K_f$. This constant represents the reverse reaction for the decomposition of a complex ion into its individual metal and ligand components. When comparing the values for K_d, the larger the value is the more unstable the complex ion is.

As a result of these complex ions forming in solutions they also can play a key role in solubility of other compounds. When a complex ion is formed it can alter the concentrations of its components in the solution. For example:

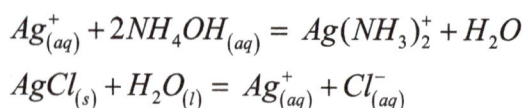

$$Ag^+_{(aq)} + 2NH_4OH_{(aq)} = Ag(NH_3)^+_2 + H_2O$$

$$AgCl_{(s)} + H_2O_{(l)} = Ag^+_{(aq)} + Cl^-_{(aq)}$$

In these reactions which both occurred in the same reaction vessel, the solubility of the silver chloride would be increased as a result of the formation of the complex ion. The complex ion formation is favorable takes away a significant portion of the silver ions in solution, as a result the equilibrium for the formation of silver ions from silver chloride will shift to the right to make up for the deficit.

This new solubility can be calculated given the values of K_f and K_{sp} for the original reactions. The solubility is found essentially by combining the two separate equilibria into one combined equilibrium reaction and this combined reaction is the one that determines the new solubility. So K_c, the new solubility constant, is denoted by $K_c = K_{sp} * K_f$.

Application of Coordination Compounds

Metals only exist in solution as coordination complexes, it follows then that this class of compounds is useful in a wide variety of ways. Coordination compounds are therefore found both in nature and in industry (in, especially, color-rich ways). Some common complex ions include such substances as vitamin B_{12}, the heme group in hemoglobin and the cytochromes, and the chlorin group in chlorophyll (which are dark red or cherry colored, blood red, and green in color respectively), and some dyes and pigments. One major use of coordination compounds is in homogeneous catalysis for the production of organic substances.

Coordination compounds have uses in both nature and in industry. Coordination compounds are vital to many living organisms. For example, many enzymes are metal complexes, like carboxypeptidase, a hydrolytic enzyme important in digestion. This enzyme consists of a zinc ion surrounded by many amino acid residues. Another complex ion enzyme is catalase, which decomposes the cell's waste hydrogen peroxide. This enzyme contains iron-porphyrin complexes, similar to those of heme in the hemoglobin molecule. Chlorophyll contains a magnesium-porphyrin complexes (chlorin), and vitamin B_{12} is a complex with cobalt and corrin.

Coordination compounds are also widely used in industry. The intense colors of many compounds render them of great use as dyes and pigments. Specifically phthalocyanine complexes are an important class of dyes for fabrics. Nickel, cobalt, and copper can be extracted using hydrometallurgical processes involving complex ions. They are extracted from their ores as ammine complexes with aqueous ammonia. Metals can also be separated using the selective precipitation and solubility of complex ions, as explained in later paragraphs. Cyanide complexes are often used in electroplating.

Coordination compounds can also be used to identify unknown substances in a solution. This analysis can be done by utilizing the selective precipitation of the complex ions, the formation of color complexes which can be measured spectrophotometrically, or the preparation of complexes, such as metal acetylacetonates, which can be separated with organic solvents.

A combination of titanium trichloride and triethylaluminum brings about the polymerization of organic compounds with carbon-carbon double bonds to form polymers of high molecular weight and ordered structures. Many of these polymers are of great commercial importance because they are used in common fibers, films, and plastics.

Other common uses of coordination compounds in industry include the following:

1. They are used in photography, i.e., AgBr forms a soluble complex with sodium thiosulfate in photography.

2. $K[Ag(CN)_2]$ is used for electroplating of silver, and $K[Au(CN)_2]$ is used for gold plating.

3. Some ligands oxidise Co^{2+} to Co^{3+} ion.

4. Ethylenediaminetetraacetic acid (EDTA) is used for estimation of Ca^{2+} and Mg^{2+} in hard water.

5. Silver and gold are extracted by treating zinc with their cyanide complexes.

Werner's Coordination Theory

In 1893, Werner was the first to propose correct structures for coordination compounds containing complex ions , in which a central transition metal atom is surrounded by neutral or anionic ligands .

$CoCl_3 \cdot 6NH_3$ This type of representation does not implies any proper geometry around the central metal ion.

But in experiment

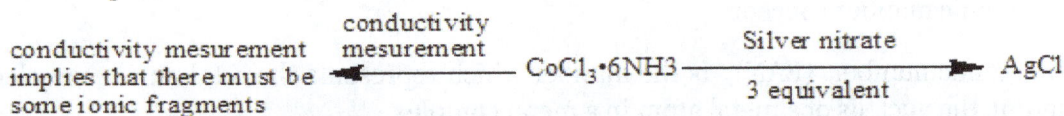

conductivity mesurement implies that there must be some ionic fragments ← conductivity mesurement — $CoCl_3 \cdot 6NH3$ → Silver nitrate / 3 equivalent → AgCl

the quantitive amount of $AgNO_3$ indicate that there must be three free Cl^- and Werner proposed the structure $[Co(NH_3)_6]Cl_3$, with tho Co^{3+} ion surrounded ny six NH_3 at the vertice of an octahedron.

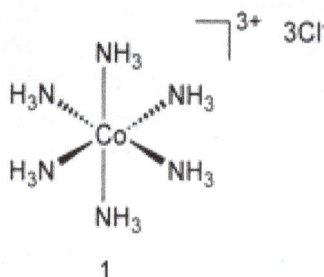

1

The basic postulates of Werner's theory may be summerised as follows:

- Metal possess two types of valency:

a) The primary or principal valency; this is the ionisable valency

e.g. In $[Co(NH_3)_5Cl]^{2+}$, Co has primary valency +3 and in $[Ni(CO)_4]$, Ni has primary valency o.

Werner's primary valence corresponds to the oxidation state.

b) A secondary nonionizable valency

- Every metal has a fixed number of secondary valency

Werner's secondary valence is always called coordination number.

e . g . In $[Co(NH_3)_5Cl]^{2+}$, Co has secondary valence equal to six and in $[Ni(CO)_4]$, Ni has secondary valence equal to 4.

- Primary valences are satisfied by negative ions whereas the secondary valences are satisfied by negative ions as well as neutral molecule or cations .

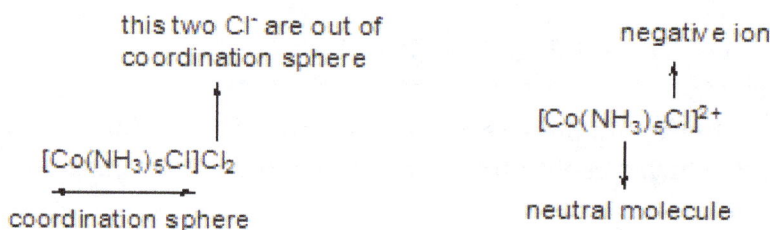

- The secondary valences are directed in space around the central metal ion in definite geometrical disposition.

Effective atomic number concept:

Effective atomic number (EAN), is the number which represents the total number of electrons surrounding the nucleus of a metal atom in a metal complex.

EAN = metal atom's electrons + the bonding electrons from the surrounding electron-donating atoms and molecules.

For case of $[Co(NH_3)_6]^{3+}$, it is 36, (trivalent cobalt ion (24) and the number of bonding electrons from six surrounding ammonia molecules, each of which contributes an electron pair).

The resulting effective atomic number is numerically equal to the atomic number of the noble-gas element found in the same period in which the metal is situated.

This rule seems to hold for most of the metal complexes with carbon monoxide , the metal carbonyls as well as many organometallic compounds (chemical compound having metal - carbon bond are known as organometallic compounds).

By using this rule it is possible to predict the number of ligands in these types of compounds and also the products of their reactions. The EAN rule is often referred to as the "18-electron rule" since, if one counts only valence electrons , the total number is 18.

18-electron rule:

Every transition metal on forming a coordination compound has a tendency to make the total number of valence electron is equal to 18. By this way the metal centre achieved pseudo noble gas configuration.

The Chelating Effect

 Crab's claw	Chelate: A compound containing a ligand (typically organic) bonded to a central metal atom at two or more points. Such crabs claw type ligand is called chelating ligand. It's also called polydentate ligand.
Types of ligands	Examples
Bidentate ligand	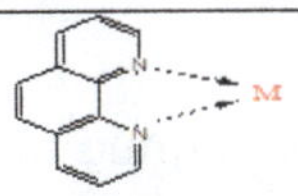 1,10-phenanthroline (o-Phen) 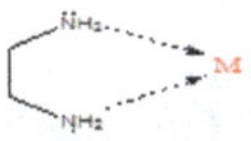 Ethylenediamine
Tridentate ligand	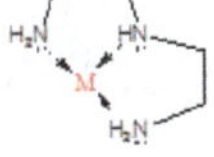 diethylenetriamine (dien)
Tetradentate ligand	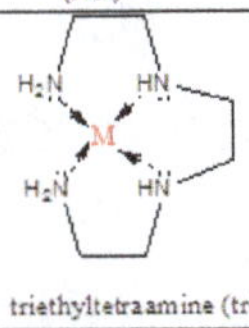 triethyltetraamine (trien)
Hexadentate ligand	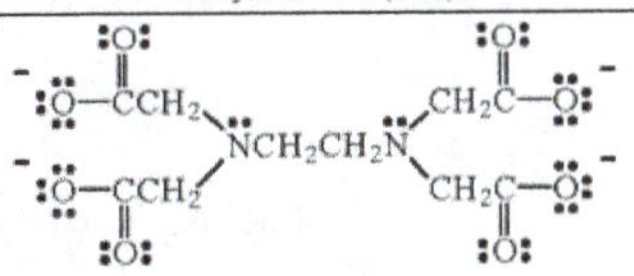 EDTA⁴⁻ This hexadentate ligand forms very stable complexes (usually octahedral structures) with most of the transition metals. The donor atoms in EDTA⁴⁻ are the two N atoms, and the four, negatively charged O atoms.

Chelates: When a multidentate ligand coordinate to a metal ion from more than one donor site forming a ring with the metal, the ligand is said to be a chelating ligand and the resulting compound is said to be a chelate complex.

Fexidentate ligand: There are many ligands which have more than one doner atoms and they are able to bond with mental center in several fasion such as monodentate fasion, bidentate fasion etc.

e.g SO_4^{2-}, CO_3^{2-} etc.

In $[Co(NH_3)_4 SO_4]^+$, SO_4^{2-} bind with the mental center as bidentate fasion and in $[Co(NH_3)_5 SO_4]^+$, SO_4^{2-} bind with the mental center as monodentate fasion.

Nomenclature of coordination compounds:

1) Orders of naming ions in coordination complexes :

The names of neutral coordination complexes are given without spaces and for ionic coordination complexes the cation is named first & then the anion separated by a space.

e.g mer-$[Ru(PPh_3)_3Cl_3]$

mer-trichlorotris(triphenylphosphine)ruthenium(III)

Here the coordination complex is neutral, so no spaces are necessary. The word ' mer' is used for expressing the geometry.

$$Na\,[PtCl_3\,(NH_3)]$$
Sodium amminetrichloroplatinate(II)

In the above examples, the cations sodium is named first and then separated by a space from the names of the anions

2) Naming the coordination sphere: ligands are named first & then the metal ion.

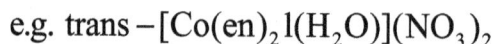

$$e.g.\ trans-[Co(en)_2 l(H_2O)](NO_3)_2$$
$$trans-Aquabis(ethylenediamine)\,iodocobalt(III)\,nitrate$$

3) Names of the ligands: Negative ligands end in $-$o & the positive ligands end in $-$ium. The neutral ligands are named as such.

$$e.g.\ K_2[CuBr_4]$$
Potassium tetrabromocuprate(II)

$$(NH_4)_2[Ni(C_2O_4)_2(H_2O)_2]$$
ammonium diaquabis(oxalato)nickelate(II)

4) Order of naming the ligands: alphabetical order irrespective of their charge. (IUPAC convention)

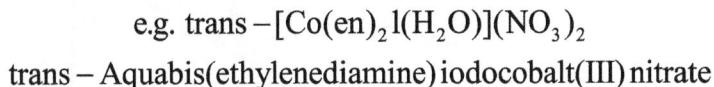

e.g. trans $-[Co(en)_2 l(H_2O)](NO_3)_2$

trans $-$ Aquabis(ethylenediamine)iodocobalt(III) nitrate

5) Numerical prefix to indicate the number of ligands:

For simple ligands							
No. of ligand/s	1	2	3	4	5	6	....
Prefixes	No need to mention mono	di	tri	tetra	penta	hexa	So on

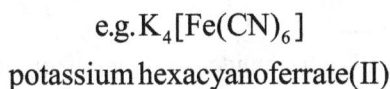

e.g. $K_4[Fe(CN)_6]$

potassium hexacyanoferrate(II)

Ligands contain the affixes di, tri, etc. e.g. ethylenediamine					
No. of ligand/s	1	2	3	4	....
Prefixes	No need to mention mono	bis	tris	tetrakis	So on

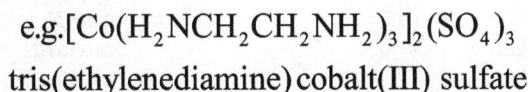

e.g. $[Co(H_2NCH_2CH_2NH_2)_3]_2(SO_4)_3$

tris(ethylenediamine) cobalt(III) sulfate

Remember that you never have to indicate the number of cations and anions in the name of an ionic compound.

6) Ending of names:

Types of complex/ coordination sphere	Central metal atom ends in
anionic	-ate
Cationic or neutral	Without any characteristic ending

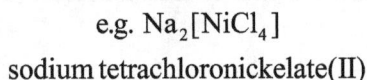

e.g. $Na_2[NiCl_4]$

sodium tetrachloronickelate(II)

And for neutral coordination complex such as $Pt(NH_3)_2Cl_4$

the IUPAC name of the coordination complex diamminetetrachloraplatinum(IV)

7) Oxidation state of the central ion: Roman numerical (such as II, III, IV) at the end of metal part without any spacing.

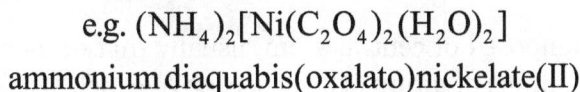

e.g. $(NH_4)_2[Ni(C_2O_4)_2(H_2O)_2]$

ammonium diaquabis(oxalato)nickelate(II)

8) Bridging group: μ- is written as a prefix of the ligand name.

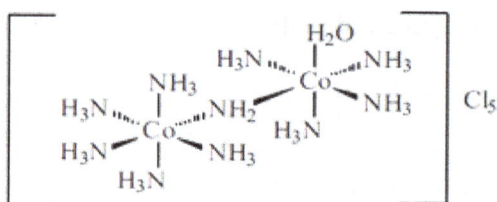

Pentamminecobalt(III) – μ – amidotetraamineaquacobalt(III) chloride

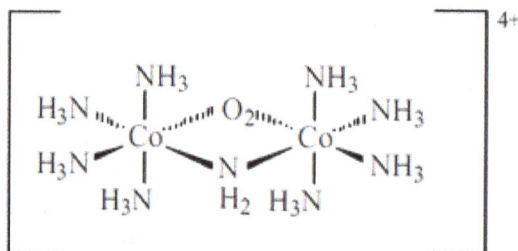

Tetraamminecobalt(III) – μ – amido – μ – superoxotetraamminecobalt(III)

9) Point of attachment: if there is more than one atom for attachment with metal then the symbol of atom is written after the name of ligand.

-SCN⁻	thiocyanato (S-thiocyanato)
-SCS⁻	isothiocyanto (N-thiocyanto)
-NCSe⁻	isoselenocyanato (N-selenocyanato)
-NO2⁻	nitrito-N
-ONO⁻	nitrito-O

e.g. $[CoCl(NO_2)(NH_3)_4]^+$
tetraamminechloronitrito-N-cobalt(III)

10) Isomer:

Type of isomer	Term use
Geometrical isomer	cis/ trans depending upon geometry
optical	+, - or d and 1 respectively

e.g. trans-$[Co(en)_2I(H_2O)](NO_3)_2$
trans-Aquabis(ethylenediamine)iodocobalt(III) nitrate

Valance Theory of Coordination Complexes

According to VBT theory, a coordination entity is formed as a result of coordinate covalent bond formation by electron pairs from ligands (Lewis bases) through overlap of appropriate atomic orbitals (usually hybrid orbitals) of the metal (Lewis acid) and ligand.

A coordination entity is composed of central atom, usually that of metal, to which is attached a surrounding array of other atoms or group of atoms, each of which is called ligand.

Hypothetical sequence of steps for the formation of a coordination entity;

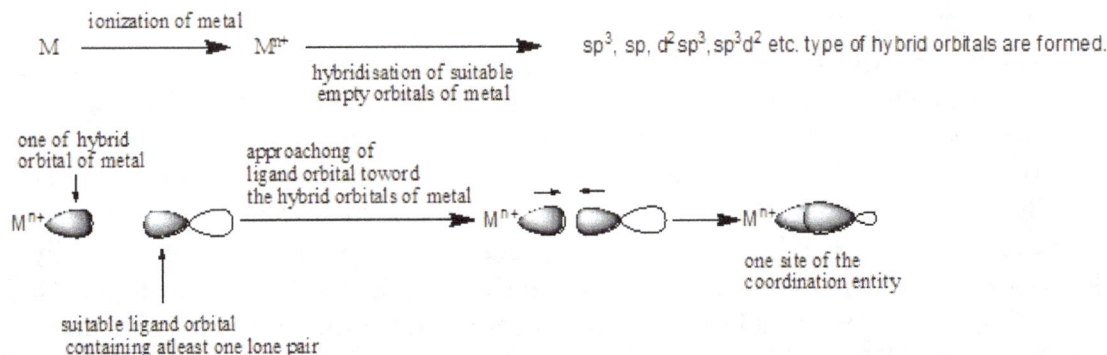

Some Hybridization schemes (δ-only) for complex compounds in common geometry			
Coordination number	Geometry	Hybridization	Examples
2	Linear	sp (s,p_z)	$[Ag(NH_3)_2]^+$
3	Trigonal planer	sp^2 (s, p_x,p_y)	$[Ag(PR_3)_3]$
4	Tetrahedral	sp^3 (s, p_x,p_y,p_z)	$[Be(H_2O)_4]^{2+}$
		d^{3}s (d_{xy},d_{yz},d_{xz},s)	MnO_4^-
	Square planer	dsp^2 $(d_{z^2}, s, p_x,p_y,)$	$[Ni(CN)_4]^{2-}$
5	Trigonal bipyramid	dsp^3 $(d_{z^2}, s, p_x,p_y,p_z)$	$[CuCl_5]^{3-}$
	Square pyramid	dsp^3 $(d_{x^2-y^2}, s, p_x,p_y,p_z)$	$[VO(acac)_2]$
6	Octahedral	d^{2}sp^3 $(d_{z^2},d_{x^2-y^2}, s, p_x,p_y,p_z)$	$[Co(NH_3)_6]^{3+}$
		sp^{3}d^2	$[CoF_6]^{3-}$
	Trigonal prism	d^{2}sp^3 $(d_{yz},d_{xz}, s, p_x,p_y,p_z)$	$[Mo(S_2C_2Ph_2)_3]$

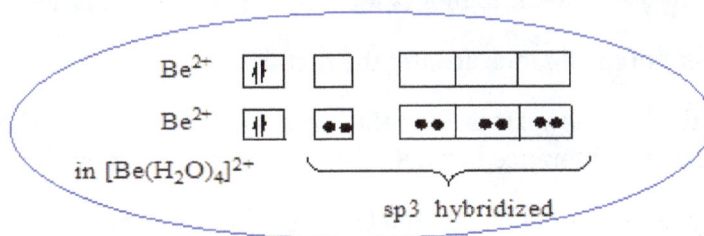

Limitation of the VBT:

1. Fail to explain the colour & characteristics of absorption spectra of complex compounds.

2. Orbital contribution and temperature dependency on magnetic moment of coordination complex are not properly explained by VBT.

3. It is not helpful to predict the mystery of formation of outer or inner orbital coordination complex.

4. VBT fails to predict any distortion in the shapes of the coordination complexes from regular geometry.

Crystal Field Theory

Crystal Field Theory (CFT) is a model that describes the breaking of degeneracies of electron orbital states, usually d or f orbitals, due to a static electric field produced by a surrounding charge distribution (anion neighbors). This theory has been used to describe various spectroscopies of transition metal coordination complexes, in particular optical spectra (colors). CFT successfully accounts for some magnetic properties, colours, hydration enthalpies, and spinel structures of transition metal complexes, but it does not attempt to describe bonding. CFT was developed by physicists Hans Bethe and John Hasbrouck van Vleck in the 1930s. CFT was subsequently combined with molecular orbital theory to form the more realistic and complex ligand field theory (LFT), which delivers insight into the process of chemical bonding in transition metal complexes.

Overview of Crystal Field Theory Analysis

According to Crystal Field Theory, the interaction between a transition metal and ligands arises from the attraction between the positively charged metal cation and negative charge on the non-bonding electrons of the ligand. The theory is developed by considering energy changes of the five degenerate d-orbitals upon being surrounded by an array of point charges consisting of the ligands. As a ligand approaches the metal ion, the electrons from the ligand will be closer to some of the d-orbitals and farther away from others, causing a loss of degeneracy. The electrons in the d-orbitals and those in the ligand repel each other due to repulsion between like charges. Thus the d-electrons closer to the ligands will have a higher energy than those further away which results in the d-orbitals splitting in energy. This splitting is affected by the following factors:

- the nature of the metal ion.

- the metal's oxidation state. A higher oxidation state leads to a larger splitting.

- the arrangement of the ligands around the metal ion.

- the nature of the ligands surrounding the metal ion. The stronger the effect of the ligands then the greater the difference between the high and low energy d groups.

The most common type of complex is octahedral; here six ligands form an octahedron around the metal ion. In octahedral symmetry the d-orbitals split into two sets with an energy difference, Δ_{oct} (the crystal-field splitting parameter) where the d_{xy}, d_{xz} and d_{yz} orbitals will be lower in energy than the d_{z^2} and $d_{x^2-y^2}$, which will have higher energy, because the former group is farther from the ligands than the latter and therefore experience less repulsion. The three lower-energy orbitals are collectively referred to as t_{2g}, and the two higher-energy orbitals as e_g. (These labels are based on the theory of molecular symmetry).

Tetrahedral complexes are the second most common type; here four ligands form a tetrahedron around the metal ion. In a tetrahedral crystal field splitting, the d-orbitals again split into two groups, with an energy difference of Δ_{tet}. The lower energy orbitals will be d_{z^2} and $d_{x^2-y^2}$, and the higher energy orbitals will be d_{xy}, d_{xz} and d_{yz} - opposite to the octahedral case. Furthermore, since the ligand electrons in tetrahedral symmetry are not oriented directly towards the d-orbitals, the energy splitting will be lower than in the octahedral case. Square planar and other complex geometries can also be described by CFT.

The size of the gap Δ between the two or more sets of orbitals depends on several factors, including the ligands and geometry of the complex. Some ligands always produce a small value of Δ, while others always give a large splitting. The reasons behind this can be explained by ligand field theory. The spectrochemical series is an empirically-derived list of ligands ordered by the size of the splitting Δ that they produce.

$I^- < Br^- < S^{2-} < SCN^-$ (S–bonded) $< Cl^- < NO_3^- < N_3^- < F^- < OH^- < C_2O_4^{2-} < H_2O < NCS^-$ (N–bonded) $< CH_3CN < py < NH_3 < en < 2,2'$-bipyridine $< phen < NO_2^- < PPh_3 < CO < CN^-$.

It is useful to note that the ligands producing the most splitting are those that can engage in metal to ligand back-bonding.

The oxidation state of the metal also contributes to the size of Δ between the high and low energy levels. As the oxidation state increases for a given metal, the magnitude of Δ increases. A V^{3+} complex will have a larger Δ than a V^{2+} complex for a given set of ligands, as the difference in charge density allows the ligands to be closer to a V^{3+} ion than to a V^{2+} ion. The smaller distance between the ligand and the metal ion results in a larger Δ, because the ligand and metal electrons are closer together and therefore repel more.

High-spin and Low-spin

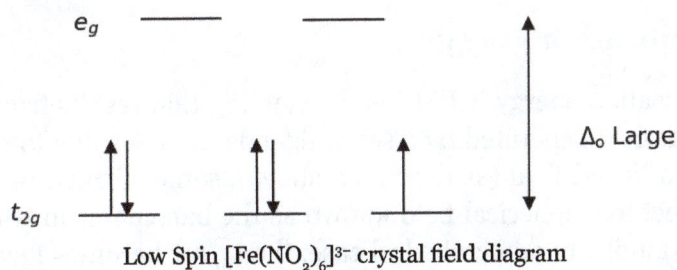

Low Spin $[Fe(NO_2)_6]^{3-}$ crystal field diagram

Ligands which cause a large splitting Δ of the d-orbitals are referred to as strong-field ligands, such as CN^- and CO from the spectrochemical series. In complexes with these ligands, it is unfavourable to put electrons into the high energy orbitals. Therefore, the lower energy orbitals are completely filled before population of the upper sets starts according to the Aufbau principle. Complexes such as this are called "low spin". For example, NO_2^- is a strong-field ligand and produces a large Δ. The octahedral ion $[Fe(NO_2)_6]^{3-}$, which has 5 d-electrons, would have the octahedral splitting diagram shown at right with all five electrons in the t_{2g} level. The low spin state therefore does not follow Hund's rule.

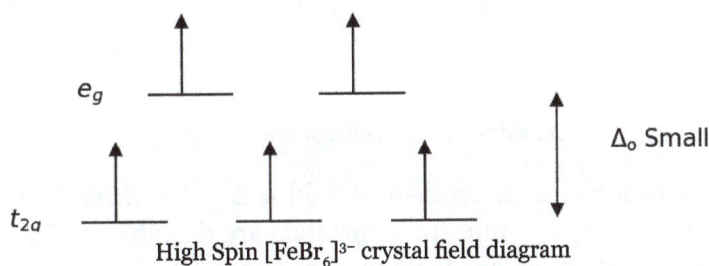

High Spin $[FeBr_6]^{3-}$ crystal field diagram

Conversely, ligands (like I⁻ and Br⁻) which cause a small splitting Δ of the d-orbitals are referred to as weak-field ligands. In this case, it is easier to put electrons into the higher energy set of orbitals than it is to put two into the same low-energy orbital, because two electrons in the same orbital repel each other. So, one electron is put into each of the five d-orbitals before any pairing occurs in accord with Hund's rule and "high spin" complexes are formed. For example, Br⁻ is a weak-field ligand and produces a small Δ_{oct}. So, the ion $[FeBr_6]^{3-}$, again with five d-electrons, would have an octahedral splitting diagram where all five orbitals are singly occupied.

In order for low spin splitting to occur, the energy cost of placing an electron into an already singly occupied orbital must be less than the cost of placing the additional electron into an e_g orbital at an energy cost of Δ. As noted above, e_g refers to the d_{z^2} and $d_{x^2-y^2}$ which are higher in energy than the t_{2g} in octahedral complexes. If the energy required to pair two electrons is greater than Δ, the energy cost of placing an electron in an e_g, high spin splitting occurs.

The crystal field splitting energy for tetrahedral metal complexes (four ligands) is referred to as Δ_{tet}, and is roughly equal to $4/9\Delta_{oct}$ (for the same metal and same ligands). Therefore, the energy required to pair two electrons is typically higher than the energy required for placing electrons in the higher energy orbitals. Thus, tetrahedral complexes are usually high-spin.

The use of these splitting diagrams can aid in the prediction of the magnetic properties of coordination compounds. A compound that has unpaired electrons in its splitting diagram will be paramagnetic and will be attracted by magnetic fields, while a compound that lacks unpaired electrons in its splitting diagram will be diamagnetic and will be weakly repelled by a magnetic field.

Crystal Field Stabilization Energy

The crystal field stabilization energy (CFSE) is the stability that results from placing a transition metal ion in the crystal field generated by a set of ligands. It arises due to the fact that when the d-orbitals are split in a ligand field (as described above), some of them become lower in energy than before with respect to a spherical field known as the barycenter in which all five d-orbitals are degenerate. For example, in an octahedral case, the t_{2g} set becomes lower in energy than the orbitals in the barycenter. As a result of this, if there are any electrons occupying these orbitals, the metal ion is more stable in the ligand field relative to the barycenter by an amount known as the CFSE. Conversely, the e_g orbitals (in the octahedral case) are higher in energy than in the barycenter, so putting electrons in these reduces the amount of CFSE.

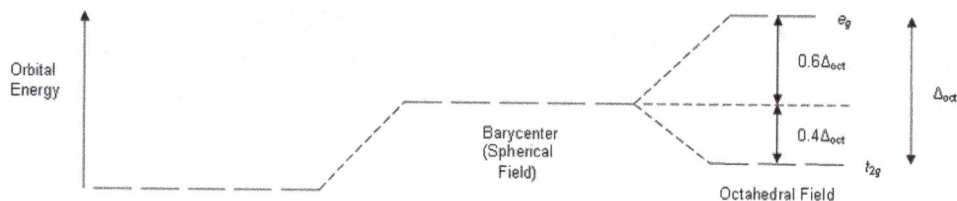

Octahedral crystal field stabilization energy

If the splitting of the d-orbitals in an octahedral field is Δ_{oct}, the three t_{2g} orbitals are stabilized relative to the barycenter by $^2/_5\,\Delta_{oct}$, and the e_g orbitals are destabilized by $^3/_5\,\Delta_{oct}$. As examples, consider the two d^5 configurations shown further up the page. The low-spin (top) example has five electrons in the t_{2g} orbitals, so the total CFSE is $5 \times \frac{2}{5}\Delta_{oct} = 2\Delta_{oct}$. In the high-spin (lower) example, the CFSE is $(3 \times \frac{2}{5}\Delta_{oct}) - (2 \times \frac{3}{5}\Delta_{oct}) = 0$ - in this case, the stabilization generated by the electrons in the lower orbitals is canceled out by the destabilizing effect of the electrons in the upper orbitals.

Optical Properties

The optical properties (details of absorption and emission spectra) of many coordination complexes can be explained by Crystal Field Theory. Often, however, the deeper colors of metal complexes arise from more intense charge-transfer excitations.

Geometries and Crystal Field Splitting Diagrams

Name	Shape	Energy diagram
Octahedral		
Pentagonal bipyramidal		
Square anti-prismatic		
Square planar		

Square pyra-midal		Square pyramidal
Tetrahedral		Tetrahedral
Trigonal bi-pyramidal		Trigonal bipyramidal

Magnetochemistry

Magnetochemistry is concerned with the magnetic properties of chemical compounds. Magnetic properties arise from the spin and orbital angular momentum of the electrons contained in a compound. Compounds are diamagnetic when they contain no unpaired electrons. Molecular compounds that contain one or more unpaired electrons are paramagnetic. The magnitude of the paramagnetism is expressed as an effective magnetic moment, μ_{eff}. For first-row transition metals the magnitude of μ_{eff} is, to a first approximation, a simple function of the number of unpaired electrons, the spin-only formula. In general, spin-orbit coupling causes μ_{eff} to deviate from the spin-only formula. For the heavier transition metals, lanthanides and actinides, spin-orbit coupling cannot be ignored. Exchange interaction can occur in clusters and infinite lattices, resulting in ferromagnetism, antiferromagnetism or ferrimagnetism depending on the relative orientations of the individual spins.

Magnetic Susceptibility

The primary measurement in magnetochemistry is magnetic susceptibility. This measures the strength of interaction on placing the substance in a magnetic field. The volume magnetic susceptibility, represented by the symbol χ_v is defined by the relationship

$$\vec{M} = \chi_v \vec{H}$$

where, $\vec{M}$ is the magnetization of the material (the magnetic dipole moment per unit volume), measured in amperes per meter (SI units), and $\vec{H}$ is the magnetic field strength, also measured in amperes per meter. Susceptibility is a dimensionless quantity. For chemical applications the molar magnetic susceptibility (χ_{mol}) is the preferred quantity. It is measured in $m^3 \cdot mol^{-1}$ (SI) or $cm^3 \cdot mol^{-1}$ (CGS) and is defined as

$$\chi_{mol} = M \chi_v / \rho$$

where ρ is the density in $kg \cdot m^{-3}$ (SI) or $g \cdot cm^{-3}$ (CGS) and M is molar mass in $kg \cdot mol^{-1}$ (SI) or $g \cdot mol^{-1}$ (CGS).

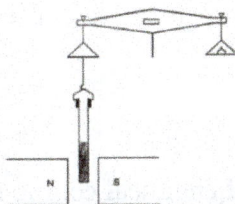

Schematic diagram of Gouy balance

A variety of methods are available for the measurement of magnetic susceptibility.

- With the Gouy balance the weight change of the sample is measured with an analytical balance when the sample is placed in a homogeneous magnetic field. The measurements are calibrated against a known standard, such as mercury cobalt thiocyanate, $HgCo(NCS)_4$. Calibration removes the need to know the density of the sample. Variable temperature measurements can be made by placing the sample in a cryostat between the pole pieces of the magnet.

- The Evans balance. is a torsion balance which uses a sample in a fixed position and a variable secondary magnet to bring the magnets back to their initial position. It, too, is calibrated against $HgCo(NCS)_4$.

- With a Faraday balance the sample is placed in a magnetic field of constant gradient, and weighed on a torsion balance. This method can yield information on magnetic anisotropy.

- SQUID is a very sensitive magnetometer.

- For substances in solution NMR may be used to measure susceptibility.

Types of Magnetic Behaviour

When an isolated atom is placed in a magnetic field there is an interaction because each electron in the atom behaves like a magnet, that is, the electron has a magnetic moment. There are two types of interaction.

1. Diamagnetism. When placed in a magnetic field the atom becomes magnetically polarized, that is, it develops an induced magnetic moment. The force of the interaction tends to push the atom out of the magnetic field. By convention diamagnetic susceptibility is given a negative sign. Very frequently diamagnetic atoms have no unpaired electrons *ie* each electron is paired with another electron in the same atomic orbital. The moments of the two electrons cancel each other out, so the atom has no net magnetic moment. However, for the ion Eu^{3+} which has six unpaired electrons, the orbital angular momentum cancels out the electron angular momentum, and this ion is diamagnetic at zero Kelvin.

2. Paramagnetism. At least one electron is not paired with another. The atom has a permanent magnetic moment. When placed into a magnetic field, the atom is attracted into the field. By convention paramagnetic susceptibility is given a positive sign.

When the atom is present in a chemical compound its magnetic behaviour is modified by its chemical environment. Measurement of the magnetic moment can give useful chemical information.

In certain crystalline materials individual magnetic moments may be aligned with each other (magnetic moment has both magnitude and direction). This gives rise to ferromagnetism, antiferromagnetism or ferrimagnetism. These are properties of the crystal as a whole, of little bearing on chemical properties.

Diamagnetism

Diamagnetism is a universal property of chemical compounds, because all chemical compounds contain electron pairs. A compound in which there are no unpaired electrons is said to be diamagnetic. The effect is weak because it depends on the magnitude of the induced magnetic moment. It depends on the number of electron pairs and the chemical nature of the atoms to which they belong. This means that the effects are additive, and a table of "diamagnetic contributions", or Pascal's constants, can be put together. With paramagnetic compounds the observed susceptibility can be adjusted by adding to it the so-called diamagnetic correction, which is the diamagnetic susceptibility calculated with the values from the table.

Paramagnetism

Mechanism and Temperature Dependence

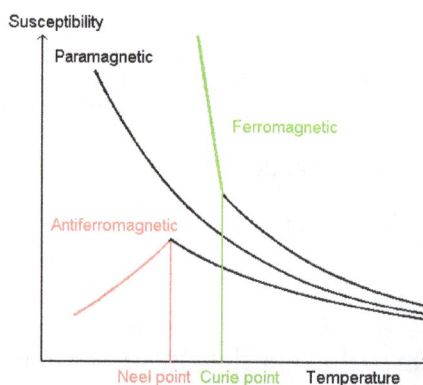

Variation of magnetic susceptibility with temperature

A metal ion with a single unpaired electron, such as Cu^{2+}, in a coordination complex provides the simplest illustration of the mechanism of paramagnetism. The individual metal ions are kept far apart by the ligands, so that there is no magnetic interaction between them. The system is said to be magnetically dilute. The magnetic dipoles of the atoms point in random directions. When a magnetic field is applied, first-order Zeeman splitting occurs. Atoms with spins aligned to the field slightly outnumber the atoms with non-aligned spins. In the first-order Zeeman effect the energy difference between the two states is proportional to the applied field strength. Denoting the energy difference as ΔE, the Boltzmann distribution gives the ratio of the two populations as , where k is the Boltzmann constant and T is the temperature in kelvins. In most cases ΔE is much smaller than kT and the exponential can be expanded as $1 - \Delta E/kT$. It follows from the presence of $1/T$ in this expression that the susceptibility is inversely proportional to temperature.

$$\chi = \frac{C}{T}$$

This is known as the Curie law and the proportionality constant, C, is known as the Curie constant, whose value, for molar susceptibility, is calculated as

$$C = \frac{Ng^2 S(S+1)\mu_B^2}{3k}$$

where N is the Avogadro constant, g is the Landé g-factor, and μ_B is the Bohr magneton. In this treatment it has been assumed that the electronic ground state is not degenerate, that the magnetic susceptibility is due only to electron spin and that only the ground state is thermally populated.

While some substances obey the Curie law, others obey the Curie-Weiss law.

$$\chi = \frac{C}{T - T_c}$$

T_c is the Curie temperature. The Curie-Weiss law will apply only when the temperature is well above the Curie temperature. At temperatures below the Curie temperature the substance may become ferromagnetic. More complicated behaviour is observed with the heavier transition elements.

Effective Magnetic Moment

When the Curie law is obeyed, the product of molar susceptibility and temperature is a constant. The effective magnetic moment, μ_{eff} is then defined as

$$\mu_{eff} = \text{constant}\sqrt{T\chi}$$

Where C has CGS units $cm^3 \; mol^{-1} \; K$, μ_{eff} is

$$\mu_{eff} = \sqrt{\frac{3k}{N\mu_B^2}}\sqrt{T\chi} \approx 2.82787\sqrt{T\chi}$$

Where C has SI units $m^3 \; mol^{-1} \; K$, μ_{eff} is

$$\mu_{eff} = \sqrt{\frac{3k}{N\mu_0\mu_B^2}}\sqrt{T\chi} \approx 797.727\sqrt{T\chi}$$

The quantity μ_{eff} is effectively dimensionless, but is often stated as in units of Bohr magneton (μ_B).

For substances that obey the Curie law, the effective magnetic moment is independent of temperature. For other substances μ_{eff} is temperature dependent, but the dependence is small if the Curie-Weiss law holds and the Curie temperature is low.

Temperature Independent Paramagnetism

Compounds which are expected to be diamagnetic may exhibit this kind of weak paramagnetism. It arises from a second-order Zeeman effect in which additional splitting, proportional to the square of the field strength, occurs. It is difficult to observe as the compound inevitably also interacts with the magnetic field in the diamagnetic sense. Nevertheless, data are available for the permanganate ion. It is easier to observe in compounds of the heavier elements, such as uranyl compounds.

Exchange Interactions

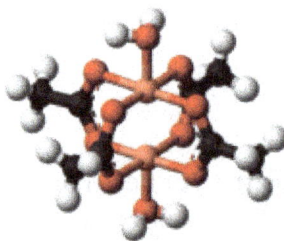

Copper(II) acetate dihydrate

Ferrimagnetic ordering in 2 dimensions

Antiferromagnetic ordering in 2 dimensions

Exchange interactions occur when the substance is not magnetically dilute and there are interactions between individual magnetic centres. One of the simplest systems to exhibit the result of exchange interactions is crystalline copper(II) acetate, $Cu_2(OAc)_4(H_2O)_2$. As the formula indicates, it contains two copper(II) ions. The Cu^{2+} ions are held together by four acetate ligands, each of which binds to both copper ions. Each Cu^{2+} ion has a d^9 electronic configuration, and so should have one unpaired electron. If there were a covalent bond between the copper ions, the electrons would pair up and the compound would be diamagnetic. Instead, there is an exchange interaction in which the spins of the unpaired electrons become partially aligned to each other. In fact two states are created, one with spins parallel and the other with spins opposed. The energy difference between the two states is so small their populations vary significantly with temperature. In consequence the magnetic moment varies with temperature in a sigmoidal pattern. The state with spins opposed has lower energy, so the interaction can be classed as antiferromagnetic in this case. It is believed that this is an example of superexchange, mediated by the oxygen and carbon atoms of the acetate ligands. Other dimers and clusters exhibit exchange behaviour.

Exchange interactions can act over infinite chains in one dimension, planes in two dimensions or over a whole crystal in three dimensions. These are examples of long-range magnetic ordering. They give rise to ferromagnetism, antiferromagnetism or ferrimagnetism, depending on the nature and relative orientations of the individual spins.

Compounds at temperatures below the Curie temperature exhibit long-range magnetic order in the form of ferromagnetism. Another critical temperature is the Néel temperature, below which antiferromagnetism occurs. The hexahydrate of nickel chloride, $NiCl_2 \cdot 6H_2O$, has a Néel temperature of 8.3 K. The susceptibility is a maximum at this temperature. Below the Néel temperature the susceptibility decreases and the substance becomes antiferromagnetic.

Complexes of Transition Metal Ions

The effective magnetic moment for a compound containing a transition metal ion with one or more unpaired electrons depends on the total orbital and spin angular momentum of the unpaired electrons, $\vec{L}$ and $\vec{S}$, respectively. "Total" in this context means "vector sum". In the approximation that the electronic states of the metal ions are determined by Russell-Saunders coupling and that spin-orbit coupling is negligible, the magnetic moment is given by

$$\mu_{eff} = \sqrt{\vec{L}(\vec{L}+1) + 4\vec{S}(\vec{S}+1)}\,\mu_B$$

Spin-only Formula

Orbital angular momentum is generated when an electron in an orbital of a degenerate set of orbitals is moved to another orbital in the set by rotation. In complexes of low symmetry certain rotations are not possible. In that case the orbital angular momentum is said to be "quenched" and is smaller than might be expected (partial quenching), or zero (complete quenching). There is complete quenching in the following cases. Note that an electron in a degenerate pair of $d_{x^2-y^2}$ or d_{z^2} orbitals cannot rotate into the other orbital because of symmetry.

	Quenched orbital angular momentum		
d^n	Octahedral		Tetrahedral
	high-spin	low-spin	
d^1			e^1
d^2			e^2
d^3	t_{2g}^3		
d^4		$t_{2g}^3 e_g^1$	
d^5		$t_{2g}^3 e_g^2$	
d^6		t_{2g}^6	$e^3 t_2^3$
d^7		$t_{2g}^6 e_g^1$	$e^4 t_2^3$
d^8	$t_{2g}^6 e_g^2$		
d^9	$t_{2g}^6 e_g^3$		

legend: t_{2g}, $t_2 = (d_{xy}, d_{xz}, d_{yz})$. e_g, $e = (d_{x^2-y^2}, d_z^2)$.

When orbital angular momentum is completely quenched $\bar{L} = 0$ and the paramagnetism can be attributed to electron spin alone. The total spin angular momentum is simply half the number of unpaired electrons and the spin-only formula results.

$$\mu_{eff} = \sqrt{n(n+2)}\mu_B$$

where n is the number of unpaired electrons. The spin-only formula is a good first approximation for high-spin complexes of first-row transition metals.

Ion	Number of unpaired electrons	Spin-only moment / μ_B	observed moment / μ_B
Ti^{3+}	1	1.73	1.73
V^{4+}	1		1.68–1.78
Cu^{2+}	1		1.70–2.20
V^{3+}	2	2.83	2.75–2.85
Ni^{2+}	2		2.8–3.5
V^{2+}	3	3.87	3.80–3.90
Cr^{3+}	3		3.70–3.90
Co^{2+}	3		4.3–5.0
Mn^{4+}	3		3.80–4.0
Cr^{2+}	4	4.90	4.75–4.90
Fe^{2+}	4		5.1–5.7
Mn^{2+}	5	5.92	5.65–6.10
Fe^{3+}	5		5.7–6.0

The small deviations from the spin-only formula may result from the neglect of orbital angular momentum or of spin-orbit coupling. For example, tetrahedral d^3, d^4, d^8 and d^9 complexes tend to show larger deviations from the spin-only formula than octahedral complexes of the same ion, because "quenching" of the orbital contribution is less effective in the tetrahedral case.

Low-spin Complexes

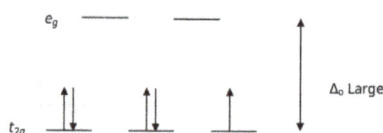

Crystal field diagram for octahedral low-spin d^5

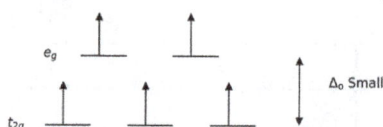

Crystal field diagram for octahedral high-spin d^5

According to crystal field theory, the d orbitals of a transition metal ion in an octahedal complex are split into two groups in a crystal field. If the splitting is large enough to overcome the energy needed to place electrons in the same orbital, with opposite spin, a low-spin complex will result.

High and low -spin octahedral complexes			
d-count	Number of unpaired electrons		examples
	high-spin	low-spin	
d^4	4	2	Cr^{2+}, Mn^{3+}
d^5	5	1	Mn^{2+}, Fe^{3+}
d^6	4	0	Fe^{2+}, Co^{3+}
d^7	3	1	Co^{2+}

With one unpaired electron μ_{eff} values range from 1.8 to 2.5 μ_B and with two unpaired electrons the range is 3.18 to 3.3 μ_B. Note that low-spin complexes of Fe^{2+} and Co^{3+} are diamagnetic. Another group of complexes that are diamagnetic are square-planar complexes of d^8 ions such as Ni^{2+} and Rh^+ and Au^{3+}.

Spin Cross-over

When the energy difference between the high-spin and low-spin states is comparable to kT (k is the Boltzmann constant and T the temperature) an equilibrium is established between the spin states, involving what have been called "electronic isomers". Tris-dithiocarbamato iron(III), $Fe(S_2CNR_2)_3$, is a well-documented example. The effective moment varies from a typical d^5 low-spin value of 2.25 μ_B at 80 K to more than 4 μ_B above 300 K.

2nd and 3rd Row Transition Metals

Crystal field splitting is larger for complexes of the heavier transition metals than for the transition metals discussed above. A consequence of this is that low-spin complexes are much more common. Spin-orbit coupling constants, ζ, are also larger and cannot be ignored, even in elementary treatments. The magnetic behaviour has been summarized, as below, together with an extensive table of data.

d-count	$kT/\zeta=0.1$ μ_{eff}	$kT/\zeta=0$ μ_{eff}	Behaviour with large spin-orbit coupling constant, ζ_{nd}
d^1	0.63	0	μ_{eff} varies with $T^{1/2}$
d^2	1.55	1.22	μ_{eff} varies with T, approximately
d^3	3.88	3.88	Independent of temperature
d^4	2.64	0	μ_{eff} varies with $T^{1/2}$
d^5	1.95	1.73	μ_{eff} varies with T, approximately

Lanthanides and Actinides

Russell-Saunders coupling, LS coupling, applies to the lanthanide ions, crystal field effects can be ignored, but spin-orbit coupling is not negligible. Consequently, spin and orbital angular momenta have to be combined,

$$\vec{L} = \sum_i \vec{l_i}$$

$$\vec{S} = \sum_i \vec{s_i}$$

$$\vec{J} = \vec{L} + \vec{S}$$

and the calculated magnetic moment is given by

$$\mu_{eff} = g\sqrt{\vec{J}(\vec{J}+1)}; g = \frac{3}{2} + \frac{\vec{S}(\vec{S}+1) - \vec{L}(\vec{L}+1)}{2\vec{J}(\vec{J}+1)}$$

Magnetic properties of trivalent lanthanide compounds														
lanthanide	Ce	Pr	Nd	Pm	Sm	Eu	Gd	Tb	Dy	Ho	Er	Tm	Yb	Lu
Number of unpaired électrons	1	2	3	4	5	6	7	6	5	4	3	2	1	0
calculated moment /μ_B	2.54	3.58	3.62	2.68	0.85	0	7.94	9.72	10.65	10.6	9.58	7.56	4.54	0
observed moment /μ_B	2.3–2.5	3.4–3.6	3.5–3.6		1.4–1.7	3.3–3.5	7.9–8.0	9.5–9.8	10.4–10.6	10.4–10.7	9.4–9.6	7.1–7.5	4.3–4.9	0

In actinides spin-orbit coupling is strong and the coupling approximates to jj coupling.

$$\vec{J} = \sum_i \vec{j_i} = \sum_i (\vec{l_i} + \vec{s_i})$$

This means that it is difficult to calculate the effective moment. For example, uranium(IV), f^2, in the complex $[UCl_6]^{2-}$ has a measured effective moment of 2.2 μ_B, which includes a contribution from temperature-independent paramagnetism.

Main Group Elements and Organic Compounds

Simulated EPR spectrum of the CH$_3$• radical

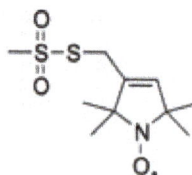

MSTL spin-label

Very few compounds of main group elements are paramagnetic. Notable examples include: oxygen, O_2; nitric oxide, NO; nitrogen dioxide, NO_2 and chlorine dioxide, ClO_2. In organic chemistry, compounds with an unpaired electron are said to be free radicals. Free radicals, with some exceptions, are short-lived because one free radical will react rapidly with another, so their magnetic properties are difficult to study. However, if the radicals are well separated from each other in a dilute solution in a solid matrix, at low temperature, they can be studied by electron paramagnetic resonance (EPR). Such radicals are generated by irradiation. Extensive EPR studies have revealed much about electron delocalization in free radicals. The simulated spectrum of the $CH_3 \cdot$ radical shows hyperfine splitting due to the interaction of the electron with the 3 equivalent hydrogen nuclei, each of which has a spin of 1/2.

Spin labels are long-lived free radicals which can be inserted into organic molecules so that they can be studied by EPR. For example, the nitroxide MTSL, a functionalized derivative of TEtra Methyl Piperidine Oxide, TEMPO, is used in site-directed spin labeling.

Applications

The gadolinium ion, Gd^{3+}, has the f^7 electronic configuration, with all spins parallel. Compounds of the Gd^{3+} ion are the most suitable for use as a contrast agent for MRI scans. The magnetic moments of gadolinium compounds are larger than those of any transition metal ion. Gadolinium is preferred to other lanthanide ions, some of which have larger effective moments, due to its having a non-degenerate electronic ground state.

For many years the nature of oxyhemoglobin, $Hb\text{-}O_2$, was highly controversial. It was found experimentally to be diamagnetic. Deoxy-hemoglobin is generally accepted to be a complex of iron in the +2 oxidation state, that is a d^6 system with a high-spin magnetic moment near to the spin-only value of 4.9 μ_B. It was proposed that the iron is oxidized and the oxygen reduced to superoxide.

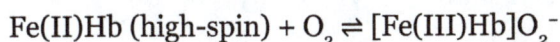

$$Fe(II)Hb \text{ (high-spin)} + O_2 \rightleftharpoons [Fe(III)Hb]O_2^-$$

Pairing up of electrons from Fe^{3+} and O_2^- was then proposed to occur via an exchange mechanism. It has now been shown that in fact the iron(II) changes from high-spin to low-spin when an oxygen molecule donates a pair of electrons to the iron. Whereas in deoxy-hemoglobin the iron atom lies above the plane of the heme, in the low-spin complex the effective ionic radius is reduced and the iron atom lies in the heme plane.

$$Fe(II)Hb + O_2 \rightleftharpoons [Fe(II)Hb]O_2 \text{ (low-spin)}$$

This information has an important bearing on research to find artificial oxygen carriers.

Compounds of gallium(II) were unknown until quite recently. As the atomic number of gallium is an odd number (31), Ga^{2+} should have an unpaired electron. It was assumed that it would act as a free radical and have a very short lifetime. The non-existence of Ga(II) compounds was part of the so-called inert pair effect. When salts of the anion with empirical formula such as $[GaCl_3]^-$ were synthesized they were found to be diamagnetic. This implied the formation of a Ga-Ga bond and a dimeric formula, $[Ga_2Cl_6]^{2-}$.

References

- Du Trémolet de Lacheisserie, Étienne; Damien Gignoux; Michel Schlenker (2005). Magnetism: Fundamentals. Springer. pp. 3–6. ISBN 0-387-22967-1

- Kirschvink, Joseph L.; Kobayashi-Kirshvink, Atsuko; Diaz-Ricci, Juan C.; Kirschvink, Steven J. (1992). "Magnetite in Human Tissues: A Mechanism for the Biological Effects of Weak ELF Magnetic Fields" (PDF). Bioelectromagnetics Supplement. 1: 101–113. Retrieved 29 March 2016

- Van Vleck, J. (1932). "Theory of the Variations in Paramagnetic Anisotropy Among Different Salts of the Iron Group". Physical Review. 41: 208. Bibcode:1932PhRv...41..208V. doi:10.1103/PhysRev.41.208

- Catherine Westbrook; Carolyn Kaut; Carolyn Kaut-Roth (1998). MRI (Magnetic Resonance Imaging) in practice (2 ed.). Wiley-Blackwell. p. 217. ISBN 0-632-04205-2

- Schmidl, Petra G. (1996–1997). "Two Early Arabic Sources On The Magnetic Compass". Journal of Arabic and Islamic Studies. 1: 81–132

- Vowles, Hugh P. (1932). "Early Evolution of Power Engineering". Isis. University of Chicago Press. 17 (2): 412–420 [419–20]. doi:10.1086/346662

- International Union of Pure and Applied Chemistry (1993). Quantities, Units and Symbols in Physical Chemistry, 2nd edition, Oxford: Blackwell Science. ISBN 0-632-03583-8. pp. 14–15

Bioinorganic Chemistry: An Overview

Bioinorganic chemistry analyses the role of metals in biology. It can be said to be an interdisciplinary subject composing of biochemistry and inorganic chemistry. It studies biological processes such as ion transport, mutation, etc. The diverse applications of bioinorganic chemistry in the current scenario have been thoroughly discussed in this chapter.

Bioinorganic Chemistry

Bioinorganic chemistry is a field that examines the role of metals in biology. Bioinorganic chemistry includes the study of both natural phenomena such as the behavior of metalloproteins as well as artificially introduced metals, including those that are non-essential, in medicine and toxicology. Many biological processes such as respiration depend upon molecules that fall within the realm of inorganic chemistry. The discipline also includes the study of inorganic models or mimics that imitate the behaviour of metalloproteins.

As a mix of biochemistry and inorganic chemistry, bioinorganic chemistry is important in elucidating the implications of electron-transfer proteins, substrate bindings and activation, atom and group transfer chemistry as well as metal properties in biological chemistry.

Composition of Living Organisms

About 99% of mammals' mass are the elements carbon, nitrogen, calcium, sodium, chlorine, potassium, hydrogen, phosphorus, oxygen and sulfur. The organic compounds (proteins, lipids and carbohydrates) contain the majority of the carbon and nitrogen and most of the oxygen and hydrogen is present as water. The entire collection of metal-containing biomolecules in a cell is called the metallome.

History

Paul Ehrlich used organoarsenic ("arsenicals") for the treatment of syphilis, demonstrating the relevance of metals, or at least metalloids, to medicine, that blossomed with Rosenberg's discovery of the anti-cancer activity of cisplatin (cis-$PtCl_2(NH_3)_2$). The first protein ever crystallized was urease, later shown to contain nickel at its active site. Vitamin B_{12}, the cure for pernicious anemia was shown crystallographically by Dorothy Crowfoot Hodgkin to consist of a cobalt in a corrin macrocycle. The Watson-Crick structure for DNA demonstrated the key structural role played by phosphate-containing polymers.

Themes in Bioinorganic Chemistry

Several distinct systems are of identifiable in bioinorganic chemistry. Major areas include:

Metal Ion Transport and Storage

This topic covers a diverse collection of ion channels, ion pumps (e.g. NaKATPase), vacuoles, siderophores, and other proteins and small molecules which control the concentration of metal ions in the cells. One issue is that many metals that are metabolically required are not readily available owing to solubility or scarcity. Organisms have developed a number of strategies for collecting such elements and transporting them.

Enzymology

Many reactions in life sciences involve water and metal ions are often at the catalytic centers (active sites) for these enzymes, i.e. these are metalloproteins. Often the reacting water is a ligand. Examples of hydrolase enzymes are carbonic anhydrase, metallophosphatases, and metalloproteinases. Bioinorganic chemists seek to understand and replicate the function of these metalloproteins.

Metal-containing electron transfer proteins are also common. They can be organized into three major classes: iron-sulfur proteins (such as rubredoxins, ferredoxins, and Rieske proteins), blue copper proteins, and cytochromes. These electron transport proteins are complementary to the non-metal electron transporters nicotinamide adenine dinucleotide (NAD) and flavin adenine dinucleotide (FAD). The nitrogen cycle make extensive use of metals for the redox interconversions.

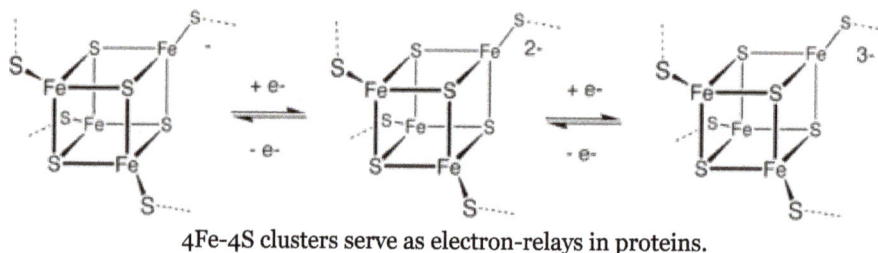

4Fe-4S clusters serve as electron-relays in proteins.

Toxicity

Several metal ions are toxic to humans and other animals. The bioinorganic chemistry of lead in the context of its toxicity has been reviewed.

Oxygen Transport and Activation Proteins

Aerobic life make extensive use of metals such as iron, copper, and manganese. Heme is utilized by red blood cells in the form of hemoglobin for oxygen transport and is perhaps the most recognized metal system in biology. Other oxygen transport systems include myoglobin, hemocyanin, and hemerythrin. Oxidases and oxygenases are metal systems found throughout nature that take advantage of oxygen to carry out important reactions such as energy generation in cytochrome c oxidase or small molecule oxidation in cytochrome P450 oxidases or methane monooxygenase. Some metalloproteins are designed to protect a biological system from the potentially harmful effects of oxygen and other reactive oxygen-containing molecules such as hydrogen peroxide. These systems include peroxidases, catalases, and superoxide dismutases. A complementary metalloprotein to those that react with oxygen is the oxygen evolving complex present in plants. This system is part of the complex protein machinery that produces oxygen as plants perform photosynthesis.

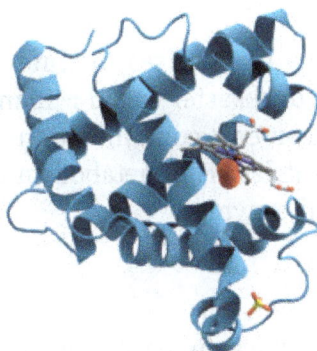

Myoglobin is a prominent subject in bioinorganic chemistry, with particular attention to the iron-heme complex that is anchored to the protein.

Bioorganometallic Chemistry

Bioorganometallic systems feature metal-carbon bonds as structural elements or as intermediates. Bioorganometallic enzymes and proteins include the hydrogenases, FeMoco in nitrogenase, and methylcobalamin. These naturally occurring organometallic compounds. This area is more focused on the utilization of metals by unicellular organisms. Bioorganometallic compounds are significant in environmental chemistry.

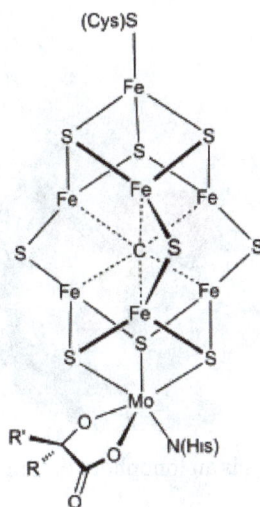

Structure of FeMoco, the catalytic center of nitrogenase.

Metals in Medicine

A number of drugs contain metals. This theme relies on the study of the design and mechanism of action of metal-containing pharmaceuticals, and compounds that interact with endogenous metal ions in enzyme active sites. The most widely used anti-cancer drug is cisplatin. MRI contrast agent commonly contain gadolinium. Lithium carbonate has been used to treat the manic phase of bipolar disorder. Gold antiarthritic drugs, e.g. auranofin have been commeriallized. Carbon monoxide-releasing molecules are metal complexes have been developed to suppress inflammation by releasing small amounts of carbon monoxide. The cardiovascular and neuronal importance of nitric oxide has been examined, including the enzyme nitric oxide synthase.

Environmental Chemistry

Environmental chemistry traditionally emphasizes the interaction of heavy metals with organisms. Methylmercury has caused major disaster called Minamata disease. Arsenic poisoning is a widespread problem owing largely to arsenic contamination of groundwater, which affects many millions of people in developing countries. The metabolism of mercury- and arsenic-containing compounds involves cobalamin-based enzymes.

Biomineralization

Biomineralization is the process by which living organisms produce minerals, often to harden or stiffen existing tissues. Such tissues are called mineralized tissues. Examples include silicates in algae and diatoms, carbonates in invertebrates, and calcium phosphates and carbonates in vertebrates.Other examples include copper, iron and gold deposits involving bacteria. Biologically-formed minerals often have special uses such as magnetic sensors in magnetotactic bacteria (Fe_3O_4), gravity sensing devices ($CaCO_3$, $CaSO_4$, $BaSO_4$) and iron storage and mobilization ($Fe_2O_3 \cdot H_2O$ in the protein ferritin). Because extracellular iron is strongly involved in inducing calcification, its control is essential in developing shells; the protein ferritin plays an important role in controlling the distribution of iron.

Types of Inorganic Elements in Biology

Alkali and Alkaline Earth Metals

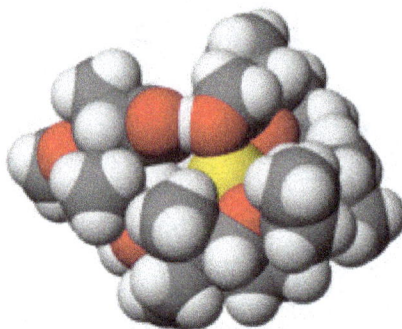

Like many antibiotics, monensin-A is an ionophore that tighlty bind Na^+ (shown in yellow).

The abundant inorganic elements act as ionic electrolytes. The most important ions are sodium, potassium, calcium, magnesium, chloride, phosphate, and the organic ion bicarbonate. The maintenance of precise gradients across cell membranes maintains osmotic pressure and pH. Ions are also critical for nerves and muscles, as action potentials in these tissues are produced by the exchange of electrolytes between the extracellular fluid and the cytosol. Electrolytes enter and leave cells through proteins in the cell membrane called ion channels. For example, muscle contraction depends upon the movement of calcium, sodium and potassium through ion channels in the cell membrane and T-tubules.

Transition Metals

The transition metals are usually present as trace elements in organisms, with zinc and iron being most abundant. These metals are used in some proteins as cofactors and are essential for the activity of enzymes such as catalase and oxygen-carrier proteins such as hemoglobin. These cofactors are bound tightly to a specific protein; although enzyme cofactors can be modified during catalysis, cofactors always return to their original state after catalysis has taken place. The metal micronutrients are taken up into organisms by specific transporters and bound to storage proteins such as ferritin or metallothionein when not being used. Cobalt is essential for the functioning of vitamin B12.

Main Group Compounds

Many other elements aside from metals are bio-active. Sulfur and phosphorus are required for all life. Phosphorus almost exclusively exists as phosphate and its various esters. Sulfur exists in a variety of oxidation states, ranging from sulfate (SO_4^{2-}) down to sulfide (S^{2-}). Selenium is a trace element involved in proteins that are antioxidants. Cadmium is important because of its toxicity.

Biological Essential Elements with their Functions

Some biological essential elements with their functions:

- Charge balance and electrolytic conductivity: Na, K, Cl

- Structure and templating: Ca, Zn, Si, S, Mo, Ni

- Signaling: Ca, B, NO

- Brønstead Acid-Base Buffering: P, Si, C

- Lewis Acid-Base Catalysis: Zn, Fe, Ni, Mn

- Electron Transfer: Fe, Cu,

- Group Transfer (e.g. CH3 , O, S): V, Fe, Co, Ni, Cu, Mo, W

- Redox Catalysis: V, Mn, Fe, Co, Ni, Cu, W, S, Se

- Energy Storage: H, P, S, Na, K, Fe

- Biomineralization: Ca, Mg, Fe, Si, Sr, Cu, P

- Energy generation: Ca, Mg

Essential amino acids:

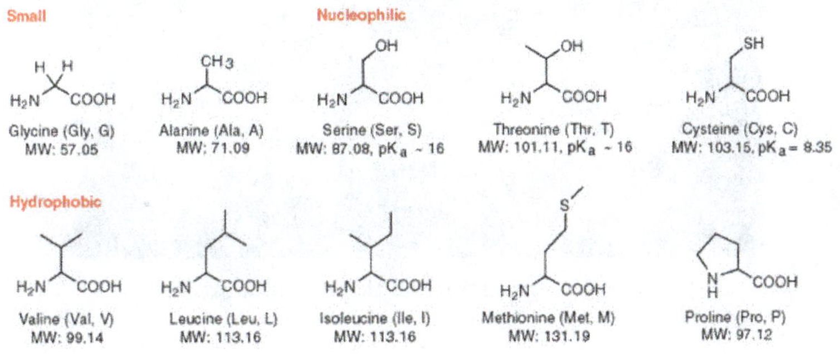

Aromatic

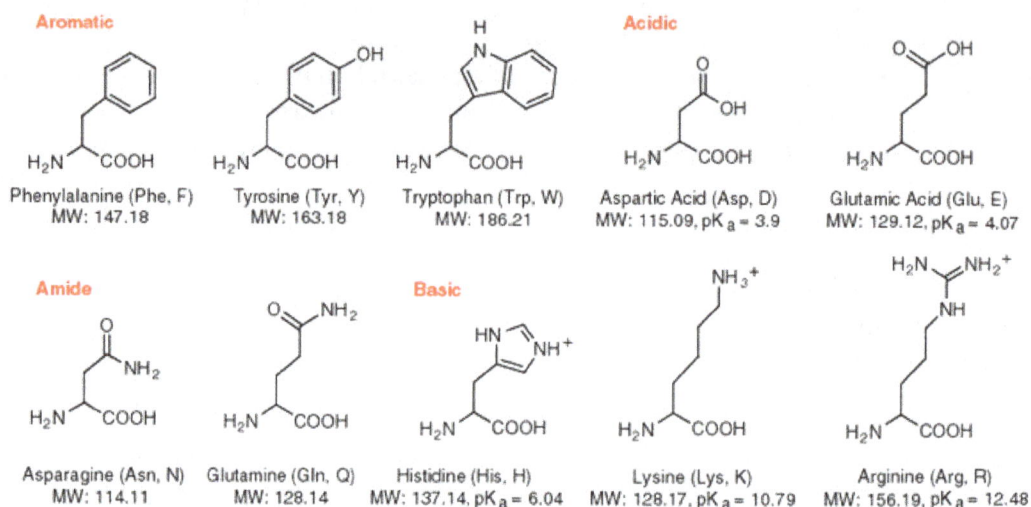

Phenylalanine (Phe, F)
MW: 147.18

Tyrosine (Tyr, Y)
MW: 163.18

Tryptophan (Trp, W)
MW: 186.21

Acidic

Aspartic Acid (Asp, D)
MW: 115.09, pK_a = 3.9

Glutamic Acid (Glu, E)
MW: 129.12, pK_a = 4.07

Amide

Basic

Asparagine (Asn, N)
MW: 114.11

Glutamine (Gln, Q)
MW: 128.14

Histidine (His, H)
MW: 137.14, pK_a = 6.04

Lysine (Lys, K)
MW: 128.17, pK_a = 10.79

Arginine (Arg, R)
MW: 156.19, pK_a = 12.48

Peptide: Polymers of monomeric amino acids are called peptides and the linkage bond between two amino acids is called peptide linkage.

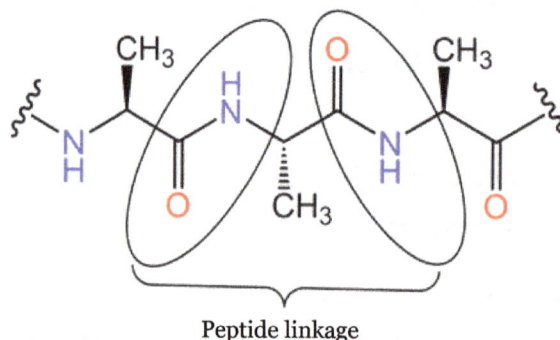

Peptide linkage

Enzyme: A biologically active compound containing one or more polypeptide units that are folded in a globular or fibrous form and catalyzes chemical reactions is called enzyme.

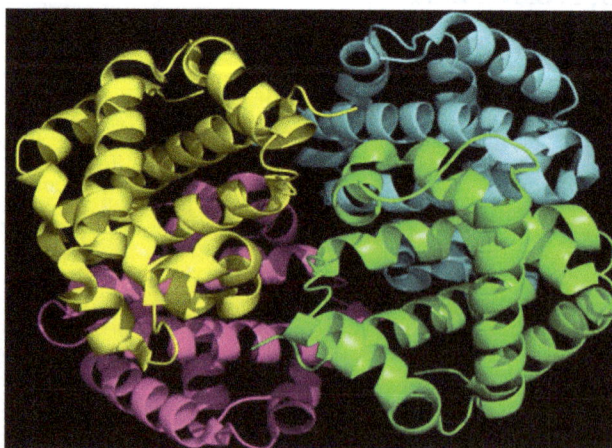

Showing folding of polypeptide chain in protein.

Apoenzyme: Many enzymes required an additional molecule to catalyze the particular chemical reaction. The small molecule is known as cofactor. It could be metal ion(s) or non-protein organic molecules. An enzyme without a cofactor is called apoenzyme.

Holoenzyme: An enzyme with a complete complement of cofactors is known as a holoenzyme.

So it can be written; Holoenzyme = Apoenzyme + coenzyme

Metalloenzyme: Enzyme that contains metal ion(s) in its active site and metal ion(s) participate(s) in the biological transformation.

Enzyme kinetics:

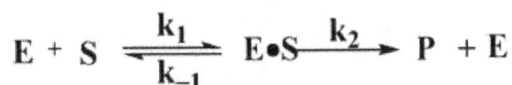

$$\text{E} + \text{S} \underset{k_{-1}}{\overset{k_1}{\rightleftharpoons}} \text{E}\bullet\text{S} \xrightarrow{k_2} \text{P} + \text{E}$$

E=enzyme, S=substrate, and P=Product

The substrate (S) binds reversibly to the enzyme (E) in the first step of an enzymatic reaction and forms E·S . After that the product is formed with release of enzyme.

The rate of E·S formation is designated by k_1 . The rates of decomposition of E·S are k_{-1} and k_2 .

At the steady state, the rate of formation of E·S will be equal to the rate of decomposition of E·S.

Hence, $k_1[\text{S}][\text{E}_{free}] = k_{-1}[\text{E}\bullet\text{S}] + k_2[\text{E}\bullet\text{S}]$, where $[\text{E}_{total}] = [\text{E}_{free}] + [\text{E}\bullet\text{S}]$

$$k_1.[\text{S}].([\text{E}_{total}] - [\text{ES}]) = k_{-1}.[\text{ES}] + k_2.[\text{ES}]$$

Solving for ES;

$$[\text{ES}] = \frac{k_1.([\text{E}_{total}][\text{S}]}{k_1.[\text{S}] + k_2 + k_{-1}} = \frac{([\text{E}_{total}][\text{S}]}{[\text{S}] + \dfrac{k_2 + k_{-1}}{k_1}}$$

The velocity of the enzyme reaction therefore is;

$$\text{Velocity} = k_2.[\text{ES}] = \frac{k_2.([\text{E}_{total}][\text{S}]}{[\text{S}] + \dfrac{k_2 + k_{-1}}{k_1}}$$

Finally, define V_{max} (the velocity at maximal concentrations of substrate) as k_2 times E_{total} , and K_M , the Michaelis-Menten constant, as $(k_2 + k_{-1})/k_1$. Substituting;

$$\text{Velocity} = V = \frac{V_{max}[S]}{[S] + K_M}$$

$$\text{Hence, } \frac{1}{V} = \frac{K_M}{V_{max}[S]} + \frac{1}{V_{max}}$$

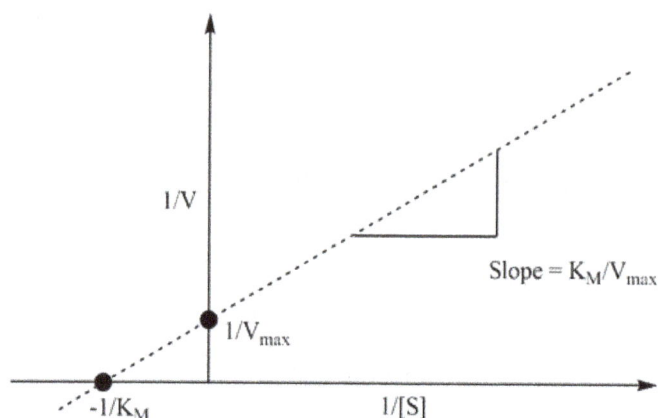

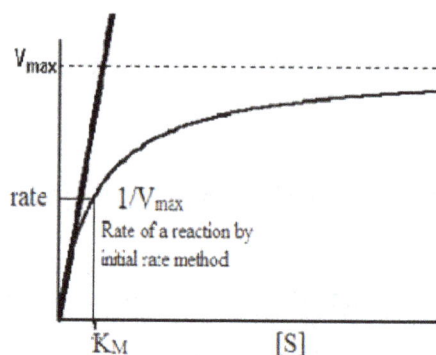

Some important facts

At low values of [S] , the initial velocity of a enzymatic reaction, V , rises almost linearly with increasing of substrate concentration, [S] .

But as the concentration of [S] increases, the gains in V level off (forming a rectangular hyperbola).

The asymptote represents the maximum velocity of the reaction, designated V_{max}

The substrate concentration that produces a V that is one-half of V_{max} is designated the Michaelis-Menten constant, K_M

K_M is (roughly) an inverse measure of the affinity or strength of binding between the enzyme and its substrate, i.e. E·S formation. The lower the K_M, the greater the affinity (so the lower the concentration of substrate needed to achieve a given rate).

Na$^+$/K$^+$-ATPase

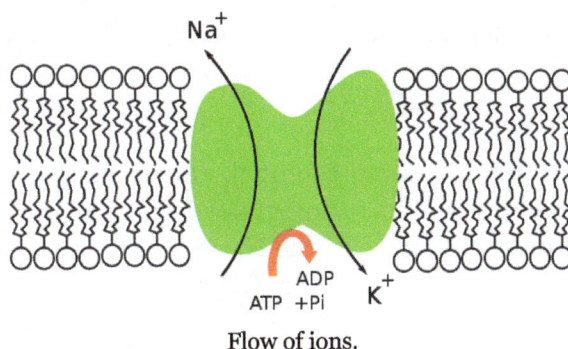

Flow of ions.

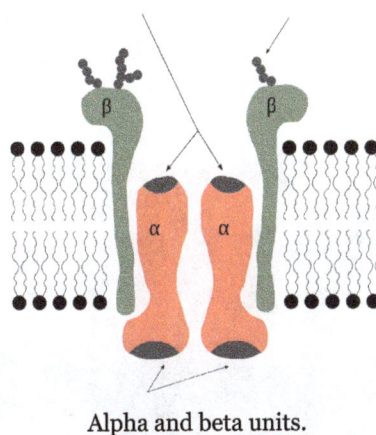

Alpha and beta units.

Na$^+$/K$^+$-ATPase (sodium-potassium adenosine triphosphatase, also known as the Na$^+$/K$^+$ pump or sodium-potassium pump) is an enzyme (EC 3.6.3.9) (an electrogenic transmembrane ATPase) found in the plasma membrane of all animal cells. The Na$^+$/K$^+$-ATPase enzyme is a solute pump that pumps sodium out of cells while pumping potassium into cells, both against their concentration gradients. This pumping is active (i.e. it uses energy from ATP) and is important for cell physiology. An example application is nerve conduction.

It has antiporter-like activity, but since it moves both molecules against their concentration gradients it is not a true antiporter, which would require one solute to move with its gradient, not against it.

Sodium-potassium Pumps

Active transport is responsible for the fact that cells contain a relatively high concentration of potassium ions but low concentrations of sodium ions. The mechanism responsible for this is the sodium-potassium pump, which moves these two ions in opposite directions across the plasma membrane. This was investigated by following the passage of radioactively labeled ions across the plasma membrane of certain cells. It was found that the concentrations of sodium and potassium ions on the two sides of the membrane are interdependent, suggesting that the same carrier transports both ions. It is now known that the carrier is an ATP-ase and that it pumps three sodium ions out of the cell for every two potassium ions pumped in.

The sodium-potassium pump was discovered in the 1950s by a Danish scientist, Jens Christian Skou, who was awarded a Nobel Prize in 1997. It marked an important step forward in the understanding of how ions get into and out of cells, and it has a particular significance for excitable cells such as nervous cells, which depend on this pump for responding to stimuli and transmitting impulses.

Function

The Na^+/K^+-ATPase helps maintain resting potential, effect transport, and regulate cellular volume. It also functions as a signal transducer/integrator to regulate MAPK pathway, ROS, as well as intracellular calcium. In most animal cells, the Na^+/K^+-ATPase is responsible for about 1/5 of the cell's energy expenditure. For neurons, the Na^+/K^+-ATPase can be responsible for up to 2/3 of the cell's energy expenditure.

Resting Potential

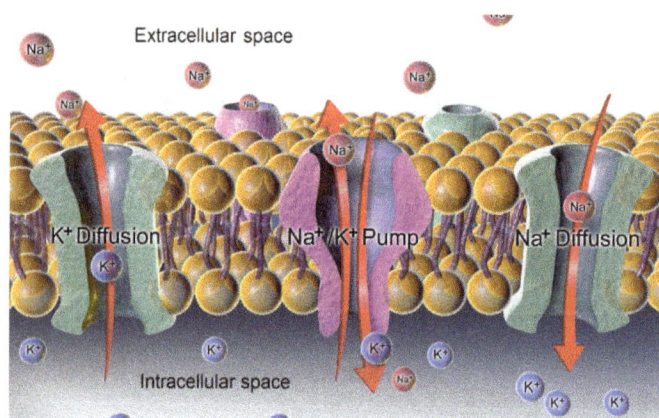

The Na^+/K^+-ATPase, as well as effects of diffusion of the involved ions maintain the resting potential across the membranes.

In order to maintain the cell membrane potential, cells keep a low concentration of sodium ions and high levels of potassium ions within the cell (intracellular). The sodium-potassium pump mechanism moves 3 sodium ions out and moves 2 potassium ions in, thus, in total, removing one positive charge carrier from the intracellular space.

It contributes -65mV to resting potential but does not in fact generate it.

Transport

Export of sodium from the cell provides the driving force for several secondary active transporters membrane transport proteins, which import glucose, amino acids, and other nutrients into the cell by use of the sodium gradient.

Another important task of the Na^+-K^+ pump is to provide a Na^+ gradient that is used by certain carrier processes. In the gut, for example, sodium is transported out of the reabsorbing cell on the blood (interstitial fluid) side via the Na^+-K^+ pump, whereas, on the reabsorbing (lumenal) side, the Na^+-glucose symporter uses the created Na^+ gradient as a source of energy to import both Na^+ and glucose, which is far more efficient than simple diffusion. Similar processes are located in the renal tubular system.

Controlling cell volume

Failure of the Na^+-K^+ pumps can result in swelling of the cell. A cell's osmolarity is the sum of the concentrations of the various ion species and many proteins and other organic compounds inside the cell. When this is higher than the osmolarity outside of the cell, water flows into the cell through osmosis. This can cause the cell to swell up and lyse. The Na^+-K^+ pump helps to maintain the right concentrations of ions. Furthermore, when the cell begins to swell, this automatically activates the Na^+-K^+ pump.

Functioning as Signal Transducer

Within the last decade [when?], many independent labs have demonstrated that, in addition to the classical ion transporting, this membrane protein can also relay extracellular ouabain-binding signalling into the cell through regulation of protein tyrosine phosphorylation. The downstream signals through ouabain-triggered protein phosphorylation events include activation of the mitogen-activated protein kinase (MAPK) signal cascades, mitochondrial reactive oxygen species (ROS) production, as well as activation of phospholipase C (PLC) and inositol triphosphate (IP3) receptor (IP3R) in different intracellular compartments.

Protein-protein interactions play a very important role in Na^+-K^+ pump-mediated signal transduction. For example, Na^+-K^+ pump interacts directly with Src, a non-receptor tyrosine kinase, to form a signaling receptor complex. Src kinase is inhibited by Na^+-K^+ pump, while, upon ouabain binding, the Src kinase domain will be released and then activated. Based on this scenario, NaKtide, a peptide Src inhibitor derived from Na^+-K^+ pump, was developed as a functional ouabain- Na^+-K^+ pump-mediated signal transduction. Na^+-K^+ pump also interacts with ankyrin, IP3R, PI3K, PLC-gamma and cofilin.

Controlling Neuron Activity States

The Na^+-K^+ pump has been shown to control and set the intrinsic activity mode of cerebellar Purkinje neurons. This suggests that the pump might not simply be a homeostatic, "housekeeping" molecule for ionic gradients; but could be a computation element in the cerebellum and the brain. Indeed, a mutation in the Na^+-K^+ pump causes rapid onset dystonia parkinsonism, which has symptoms to indicate that it is a pathology of cerebellar computation. Furthermore, an ouabain block of Na^+-K^+ pumps in the cerebellum of a live mouse results in it displaying ataxia and dystonia. Alcohol inhibits sodium-potassium pumps in the cerebellum and this is likely how it corrupts cerebellar computation and body co-ordination. The distribution of the Na^+-K^+ pump on myelinated axons, in human brain, was demonstrated to be along the internodal axolemma, and not within the nodal axolemma as previously thought.

Mechanism

- The pump, after binding ATP, binds 3 intracellular Na^+ ions.

- ATP is hydrolyzed, leading to phosphorylation of the pump at a highly conserved aspartate residue and subsequent release of ADP.

- A conformational change in the pump exposes the Na^+ ions to the outside. The phosphorylated form of the pump has a low affinity for Na^+ ions, so they are released.

- The pump binds 2 extracellular K^+ions. This causes the dephosphorylation of the pump, reverting it to its previous conformational state, transporting the K^+ ions into the cell.

- The unphosphorylated form of the pump has a higher affinity for Na^+ions than K^+ions, so the two bound K^+ions are released. ATP binds, and the process starts again.

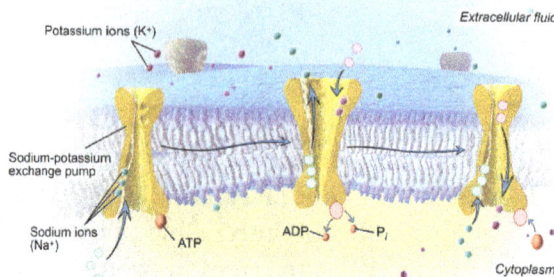

Potassium ions (K$^+$)

Extracellular fluid

Sodium-potassium
exchange pump

Sodium ions
(Na$^+$)

ATP

ADP

P$_i$

Cytoplasm

The Sodium-Potassium Exchange Pump

Regulation

Endogenous

The Na^+/K^+-ATPase is upregulated by cAMP. Thus, substances causing an increase in cAMP up-regulate the Na^+/K^+-ATPase. These include the ligands of the G_s-coupled GPCRs. In contrast, substances causing a decrease in cAMP downregulate the Na^+/K^+-ATPase. These include the ligands of the G_i-coupled GPCRs.

Note: Early studies indicated the *opposite* effect, but these were later found to be inaccurate due to additional complicating factors.

Exogenous

The Na^+-K^+-ATPase can be pharmacologically modified by administrating drugs exogenously.

For instance, Na^+-K^+-ATPase found in the membrane of heart cells is an important target of cardiac glycosides (for example digoxin and ouabain), inotropic drugs used to improve heart performance by increasing its force of contraction.

Muscle contraction is dependent on a 100- to 10,000-times-higher-than-resting intracellular Ca^{2+} concentration, which is caused by Ca^{2+} release from the muscle cells' sarcoplasmic reticulum. Immediately after muscle contraction, intracellular Ca^{2+} is quickly returned to its normal concentration by a carrier enzyme in the plasma membrane, and a calcium pump in sarcoplasmic reticulum, causing the muscle to relax.

Since this carrier enzyme (Na^+-Ca^{2+} translocator) uses the Na gradient generated by the Na^+-K^+ pump to remove Ca^{2+} from the intracellular space, slowing down the Na^+ -K^+ pump results in a permanently elevated Ca^{2+} level in the muscle, which may be the mechanism of the long-term inotropic effect of cardiac glycosides such as digoxin.

Discovery

Na+/K+-ATPase was discovered by Jens Christian Skou in 1957 while working as assistant professor at the Department of Physiology, University of Aarhus, Denmark. He published his work that year.

In 1997, he received one-half of the Nobel Prize in Chemistry "for the first discovery of an ion-transporting enzyme, Na+, K+-ATPase."

The Na+/K+ pump

The Na+/K+ pumping between inside and outside a cell is assisted by a membrane-bound Na+/K+ - ATPase enzyme that catalyzes the movement of ions in the both direction across a cell. Through this pumping process the concentration difference of the ions inside and outside a cell is maintained and hence, a constant cell potential achieved.

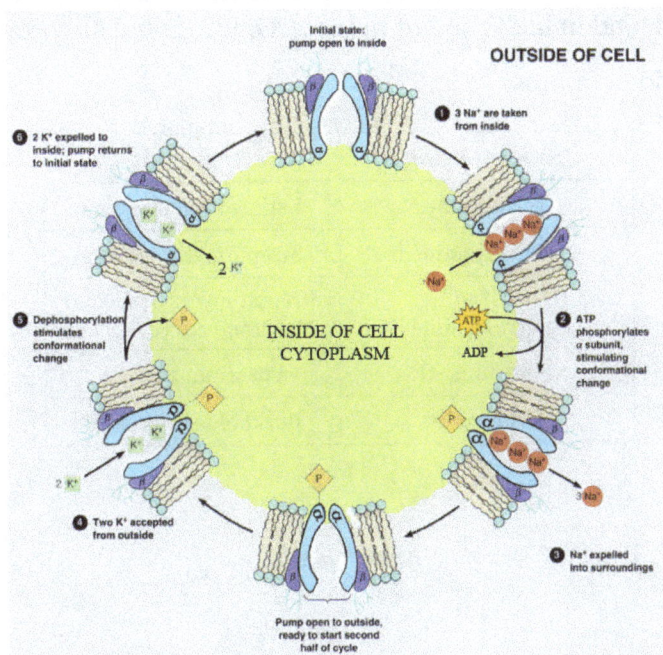

A general principle for Na+/K+ pumping carried out by Na+/K+ - ATPase enzyme.

The Na+/K+ - ATPase enzyme is a tetrameric protein with two α- subunits and two β- subunits. The α- sub units are close and contact to each other, while β- subunits are situated apart from each other (Figure). The ions exchange proceeds via following steps;

Step 1. Inside the cell three Na+ ions bind to the α- subunits of the enzyme which is consisting of six oxide units as binding site.

Step 2. The attachment of Na+ ions to the α- subunits changes the local polarity that helps an ATP to bind with a α- subunit. The bound ATP is hydrolyzed to ADP and acovalently bound with a phosphate ester (P). Due to this phosphorylation a conformation change (the inside cavity is closed and outside cavity is opened) took place in the α- subunit.

Step 3. The weakly bound three Na+ ions are released outside the cell.

Step 4 and Step 5. Two K⁺ ions outside the cell bind to the open cell cavity and simultaneously dephosphorylation takes place. Because of this a configurational change happens and two K+ ions move inside the cell with the help of a - subunits.

Step 6. The bound two K⁺ ions are released inside the cell and initial configuration is achieved by the enzyme.

Use: In the pumping process three positively charged Na⁺ ions are released from the cell and two positively charged K⁺ ions are accepted inside the cell. Hence, a charge difference is produced by the process. This charge difference creates potential gradient across the cell. For good example, signal transmission across neuron cells.

Siderophores:

Siderophores means iron carriers. They are small polydentate ligands that have high binding affinity towards iron. They found in high content in bacteria, fungi, and also grasses.

A. Hydroxamate siderophores:

Siderophore	Organism
Ferrchrome	Ustilago sphaerogena
Desferrioxamine B	Streptomyces pilosus
(Deferoxamine) Desferrioxamine E	Strephtomyces coelicolor Streptomyces coelicolor
Fusarinine C	Fusarium roseum
Ornibactin	Burkholderia cepacia

Desferrioxamine B

B. Catecholate siderophores:

Siderophore	Organism
Enterobactin	Escherichia coli
	enteric bacteria
Bacillibactin	Bacillus subtilis
	Bacillus anthracis
Vibriobactin	Vibrio cholerae

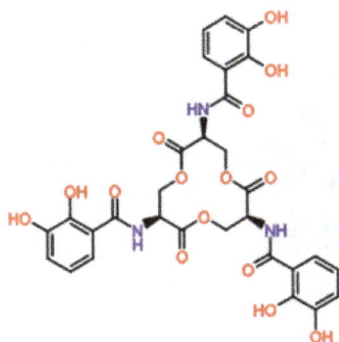

enterobactin of *E. coli*

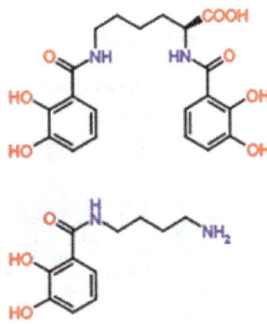

azotochelin (top) and aminochelin (bottom) of *A. vinelandii*

C. Mixed ligands:

Siderophore	Organism
azotobactin	Azotobacter vinelandii
pyoverdin	Pseudomonas aeruginosa
yersiniabactin	yersinia pestis

Ionophores:

Ionophores are lipid soluble molecules. They transport ions across the lipid bilayers of a cell membrane.

Example: Synthetic ionophores are crown ethers, cryptands, calixarenes, etc.

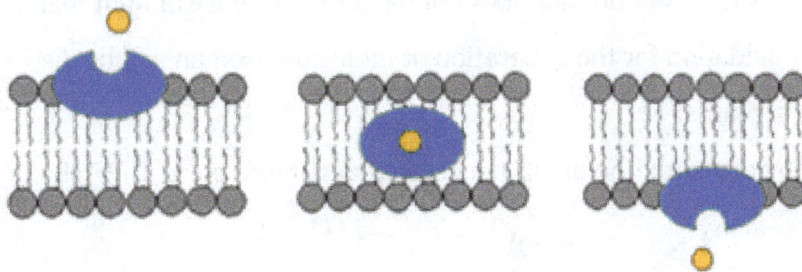

Mechanism of charge transfer.

Photosystems

The thylakoid membrane consists of both photosystem I (PS I) and photosystem II (PS II). Inside the thylokoid membrane it is called thilokoid lumen and outside is called chloroplast stroma.

chloroplast stroma

thylakoid lumen

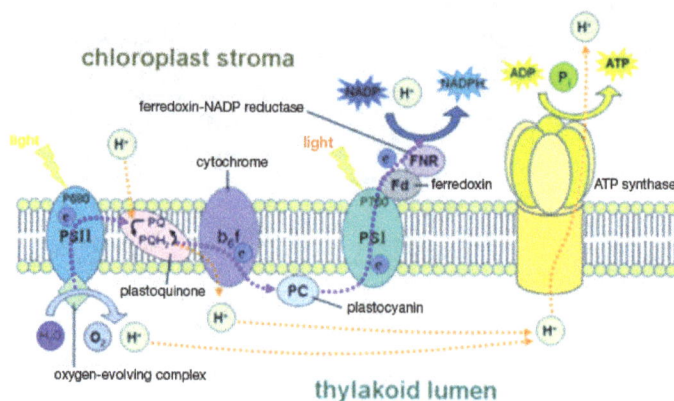

Representation of photosystem present in thilokoid membrane of chloroplast.

PS I optimally absorbs photons of a wavelength of 700 nm, and because of the wavelength it is denoted as P700. While, PS II optimally absorbs photons of a wavelength of 680 nm and hence, known as P680. PS II uses light energy to oxidize two molecules of water into one molecule of molecular oxygen. The 4 electrons removed from the water molecules are transferred through plasto quinine and cytochrome bf to PS I ultimately reduce 2NADP+ to 2NADPH. This electron transport process generates a proton gradient across the thylakoid membrane. This proton motive force is then used to drive the synthesis of ATP from ADP.

The overall reaction;

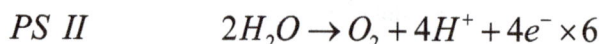

$$PS \; II \qquad 2H_2O \rightarrow O_2 + 4H^+ + 4e^- \times 6$$

$$PS \; I \qquad 2NADP^+ + 2H^+ + 4e^- \rightarrow 2NADPH \times 6$$

$$Dark \; \mathrm{Re}\, action \; 12NADPH + 12H^+ + 6CO_2 \rightarrow 12NADP^+ + 6H_2O + (CH_2O)_6$$

$$Net \; \mathrm{Re}\, action \; 6H_2O + 6CO_2 \rightarrow 6O_2 + (CH_2O)_6$$

Dark reaction means the reaction that does not need the presence of light mandatorily.

Herein, the water oxidation for the generation of molecular oxygen will be discusses.

Production of O_2 in plants:

Production of O_2 by the oxidation of H_2O in green plants takes place in PS II.

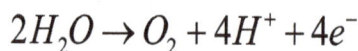

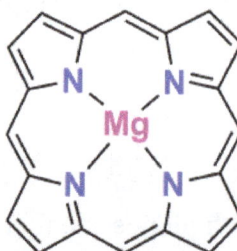

$$2H_2O \rightarrow O_2 + 4H^+ + 4e^-$$

Simplest representation of P680.

P680 contains Mg^{2+} coordinated to a prophyrin unit. It absorbs light of 680 nm wavelength and goes to its excited state (P680*). It transfer an electron to pheophytin α and becomes P680+ which is a strong oxidizing agent. It oxidizes water to oxygen where water molecule is bound to a tetra-nuclear Mn_4 cluster.

One of the proposed mechanism is shown below,

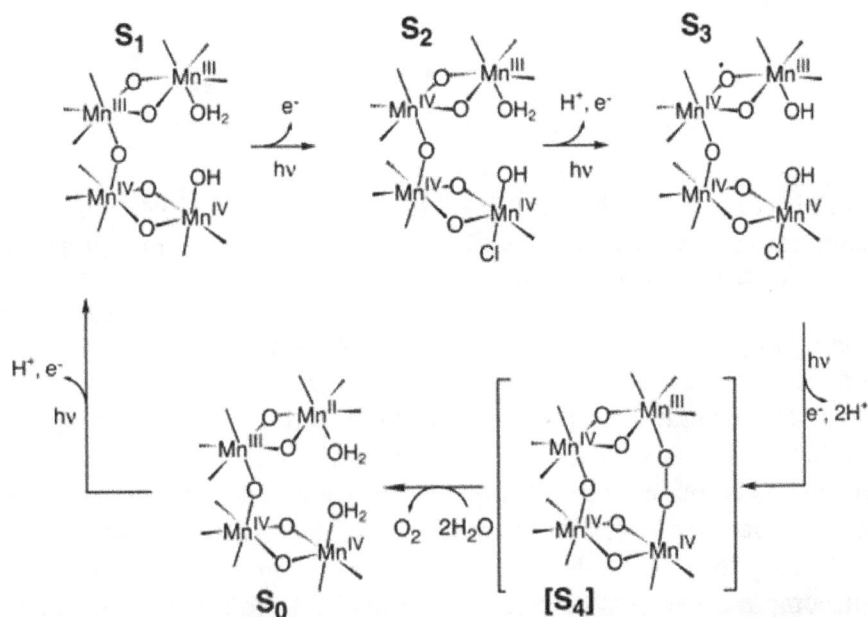

Iron Proteins:

Iron containing proteins are mainly belonging to four categories.

A. Iron-porphyrin proteins: hemoglobin, myoglobin, cytochrome P450, etc; all of them contains one of more iron-porphyrin units. They are mainly involved in oxygen transfer, oxygen storage, and electron transfer.

B. Non-heme iron proteins: ferritin, transferrin, hemesiderin, etc; they are mostly involved in iron storage and transport

C. Non-heme diiron oxo-bridged species: ribonucleotide reductase, hemerythrin, methane monooxygenase, etc.

D. Non-heme iron-sulfur cluster proteins: nitrogenase, ferredoxins, rubredoxins. They are mainly involved in biological electron transfer reactions.

Here in, some of the important iron proteins will be discussed.

Hemoglobin and Myoglobin:

Hemoglobin is an essential iron protein for molecular oxygen (dioxygen) transport and found in red blood cells. Both are globular proteins. Myoglobin is engaged in storage of molecular oxygen in muscle tissues and controlled transport of molecular oxygen for the oxidative reactions.

Hemoglobin

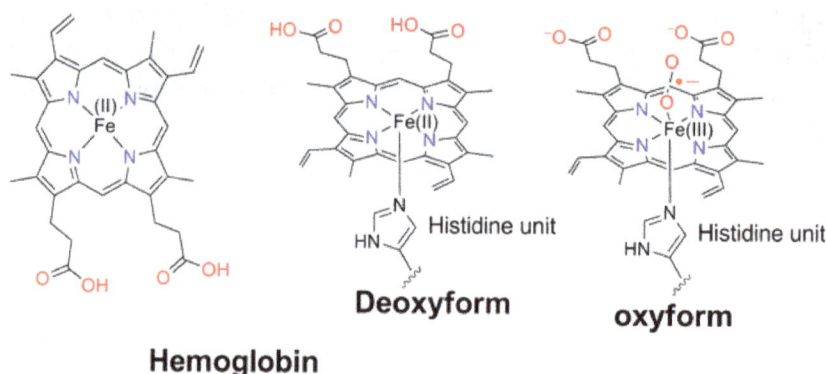

A hemoglobin unit is composed of has four protein chains each of which contains one porphyrin ring coordinated to iron (known as heme) packed in a roughly tetrahedral α2β2 cluster. The a unit contains 141 amino acids residue and the b unit contains 146 amino acids residue.

In hemoglobin, a high-spin Fe(II) is coordinated to four N atoms of porphyrin ring. The fifth coordination site is occupied by a histidine group. In this condition the protein containing four heme compartments is in stain condition and called tense (T) state. After binding to molecular oxygen high-spin Fe(III) changes to low-spin Fe(III) and molecular oxygen is transferred to superoxide. Covalent radii of Fe(II) is too large to fit into the cavity created by four N atoms of the porphyrin ring. Formation of low-spin Fe(III), the radii of iron decreases. Moreover, transformation of square pyramidal deoxy form to octahedral oxy form leads ion center closer to heme cavity. The protein is now in comparatively less stain relaxed form (R). In this was all the four iron centers in hemoglobin are transferred to oxo-hemoglobin forms.

Binding of molecular oxygen to the iron center of the hemoglobin tetramer causes release of protons from the acid units which minimize the pH. This lowering of pH favors molecular oxygen release to tissues and conversion of Fe(III) to Fe(II). This deoxyhemoglobin picks up 2 protons and 2 molecules of CO_2 form tissues and carried to the lungs, where the CO_2 is released. After that, deoxyhemoglobin which further binds to molecular oxygen and the O_2 carrying and CO_2 returning processes from tissues continue.

Myoglobin is a single chain heme protein containing 154 amino acids and it contains several region of α–helix. The structure of the active site (where the reaction occurs) and the oxygen carrying mechanism is same as hemoglobin. The oxygen uptake capacity of myoglobin is thermodynamically more compared to hemoglobin and hence, oxygen uptake in myglobin is more than that of hemoglobin.

Cytochrome c Oxidase

Cytochrome c oxidase is an electron transfer protein found at the terminal position in the respiratory transport chain of animals, plants, aerobic yeasts and in some bacteria. It reduces molecular oxygen (O_2) to H_2O.

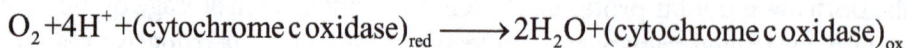

$$O_2 + 4H^+ + (\text{cytochrome c oxidase})_{red} \longrightarrow 2H_2O + (\text{cytochrome c oxidase})_{ox}$$

Active site:

The active site (where the reaction takes place) of cytochrome c oxidase is composed of a heme a_3 and Cu_B which are ~ 4.5 Å apart from each other

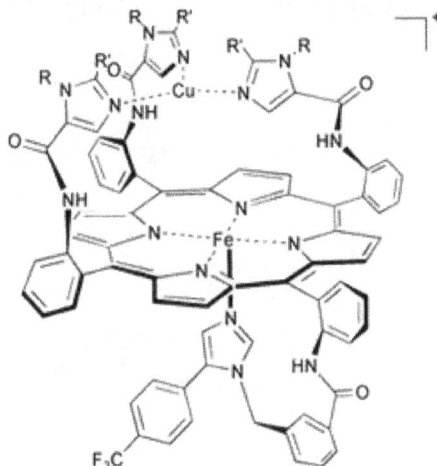

The active site of cyrochrome c oxidase.

Proposed mechanism:

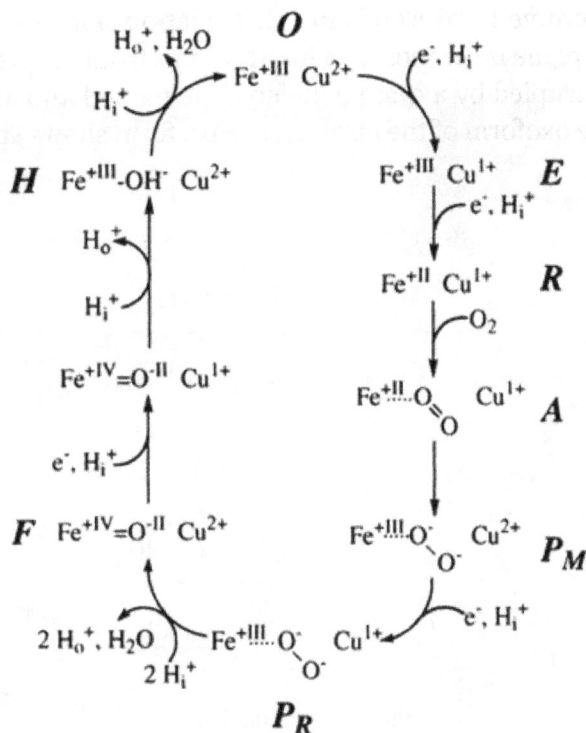

Proposed mechanism for the conversion of O_2 to H_2O by cytochrome c oxidase.

Cytochrome P450:

Heme cytochrome P450 is a large and diverse group of enzymes. The function of most cytochrome P450 enzymes is to catalyze the oxidation of organic substances.

The most common reaction catalyzed by cytochromes P450 is a monooxygenase reaction, e.g., insertion of one atom of oxygen into an organic substrate (RH) while the other oxygen atom is reduced to water:

$$RH + O_2 + NADPH + H^+ \longrightarrow ROH + H_2O + NADP^+$$

Active site:

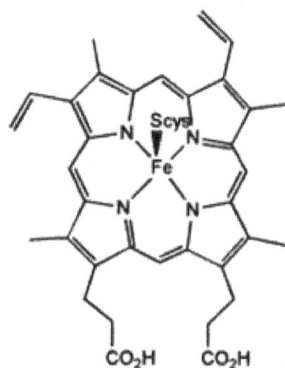

Cytochrome P450

In the active site of cytochrome P450 iron is in +III oxidation state and it is in square pyramidal geometry where the basal plane is composed of four N atoms from the prophyrin ring and the fifth, i.e. the axial position is occupied by a cystine moiety. The name P450 arises from the position of the absorption band of the oxoform of the enzyme, i.e. oxo form shows absorption band at 450 nm in the UV-Vis spectrum.

Proposed Mechanism:

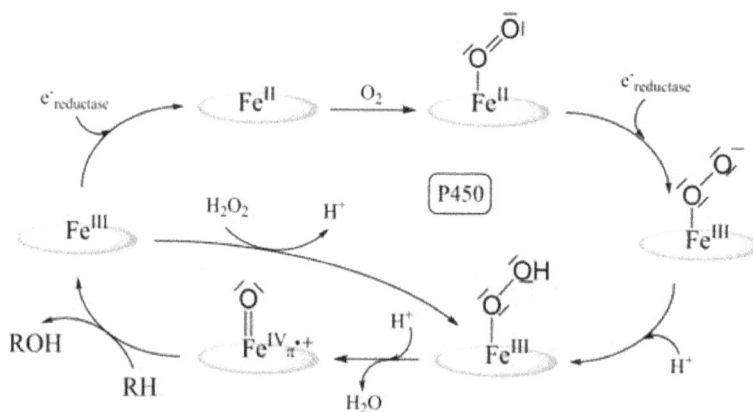

Proposed mechanism.

Catalase:

It is also a heme protein. It catalyzes disproportion of hydeogen peroxide to water and oxygen.

$$2H_2O_2 \longrightarrow H_2O + O_2$$

Active site:

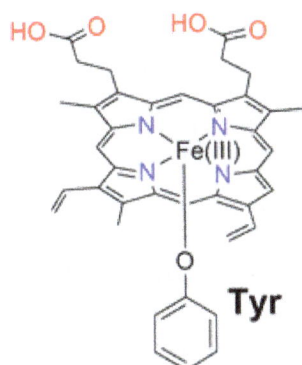

In the active site of catalase iron is in +III oxidation state and it is in square pyramidal geometry where the basal plane is composed of four N atoms from the prophyrin ring and the fifth, i.e. the axial position is occupied by a tyrosine moiety.

Proposed mechanism:

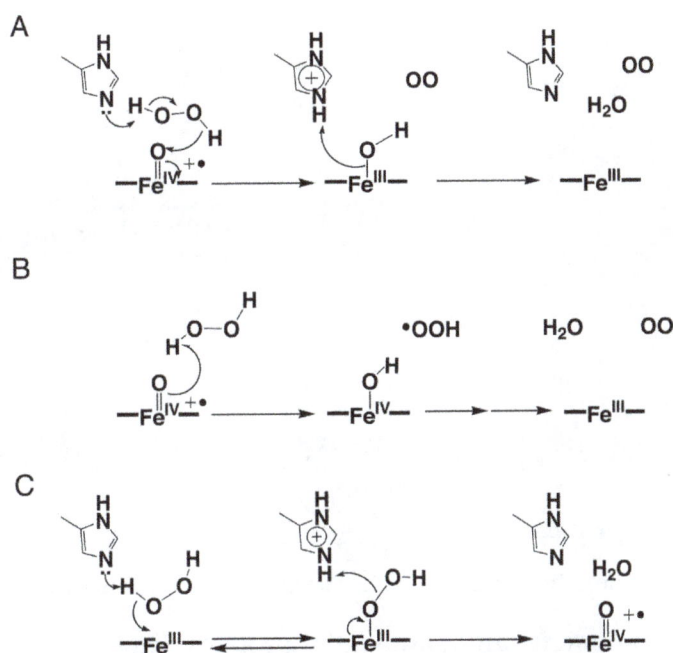

Proposed mechanisms for the catalatic reaction.
A , ionic mechanism by utilizing a general acid-base catalyst. B , radical mechanism. C , role of a general acid-base catalyst on the formation of compound I.

Iron-sulfur Proteins and Nitrogenase

The iron-sulfur proteins occurs extensively in all living organisms and take part in a wide range of electron-transfer processes, either as redox centers (e.g. ferredoxins, rubredoxins) or as catalysts (e . g . hydrogenase, nitrogenase, etc).

[1Fe-0S] proteins:

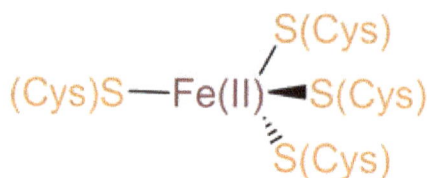

$$(Cys)S\!-\!Fe(II)\!\!\overset{S(Cys)}{\underset{S(Cys)}{\longleftarrow}}\!\!S(Cys)$$

Iron-sulfur proteins with no bridging sulfur (0S) or sulfide atom is known as rubredoxins. It is mainly found in bacteria and acts as one electron donor-acceptor. The arrangement around the iron center is tetrahedral and the Fe(II) center is surrounded by four sulfur atoms from four cystine (Cys) moieties.

[2Fe-2S], [3Fe-4S], and [4Fe-4S]:

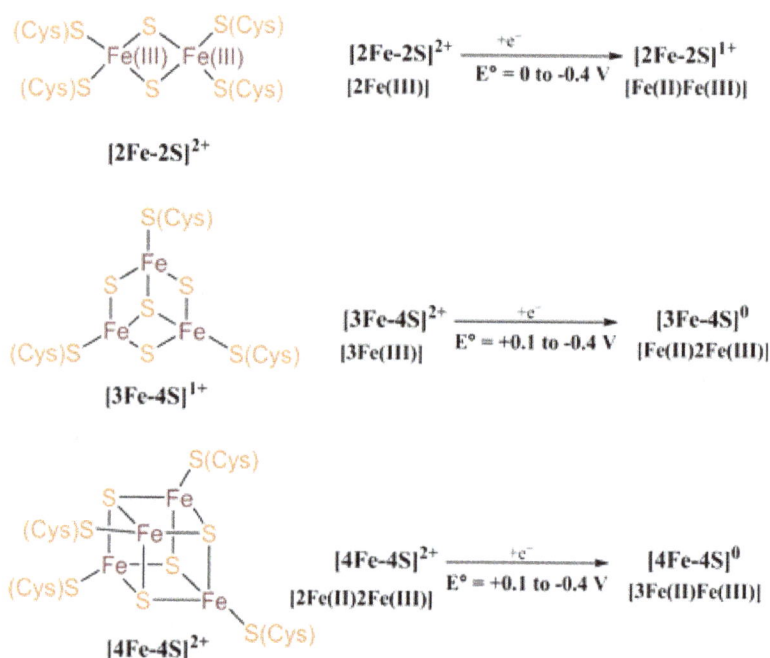

$$[2Fe\text{-}2S]^{2+} \xrightarrow[E°=0\text{ to }-0.4\text{ V}]{+e^-} [2Fe\text{-}2S]^{1+}$$
$$[2Fe(III)] \qquad\qquad\qquad [Fe(II)Fe(III)]$$

$$[3Fe\text{-}4S]^{2+} \xrightarrow[E°=+0.1\text{ to }-0.4\text{ V}]{+e^-} [3Fe\text{-}4S]^{0}$$
$$[3Fe(III)] \qquad\qquad\qquad [Fe(II)2Fe(III)]$$

$$[4Fe\text{-}4S]^{2+} \xrightarrow[E°=+0.1\text{ to }-0.4\text{ V}]{+e^-} [4Fe\text{-}4S]^{0}$$
$$[2Fe(II)2Fe(III)] \qquad\qquad\qquad [3Fe(II)Fe(III)]$$

Schematic representation for rubredoxins and ferredoxins.

Ferredoxins are most important family of iron-sulfur proteins. Three major categories of ferredoxins are, [2Fe-2S], [3Fe-4S], and [4Fe-4S]. The [4Fe-4S] is most important.

[2Fe-2S]: Isolated from mammals, plants, and bacteria. Both the iron centers are in tetrahedral coordination environment and linked by two inorganic sulfide bridges. Both Fe(III) centers are antiferromagnetically coupled to each other. Hence, a diamagnetic ground state results. After accepting a electron one center becomes Fe(II) and the other is Fe(III). After antiferromagnetic coupling between the S = ½ ground state appeared.

[3Fe-4S]: Found in Azobacter vinlandii , and Desulfovibrio gigas and also in pig heart. All the iron centers are in tetrahedral coordination environment and linked with each other by two inorganic sulfide bridges.

[4Fe-4S]: It is found is several iron containing metalloenzymes like nitrogenase, hydrogenase, etc. The structure is a distorted cubic core. The alternative corners of the cube are occupied by iron and inorganic sulfide. Irons are connected to each other via two sulfide bridges. All the iron centers are tetrahedral.

Nitrogenase:

Nitrogen-fixing bacteria found in soil and root nodules of certain plants contain an enzyme called nitrogenase that reduces molecular nitrogen to ammonia.

Reaction:

$$N_2 + 8H^+ + 8e^- + 16ATP \rightarrow 2NH_3 + H_2 + 16HPO_4^{2-}$$

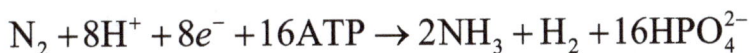

The enzyme is composed of three compartments;

A. Two ATP bound one $[Fe_4S_4]^+$ cluster

B. One P N cluster. It is a edge sharing fused structure of two $[Fe_4S_4]^+$ clusters, and

C. Fe-Mo cluster.

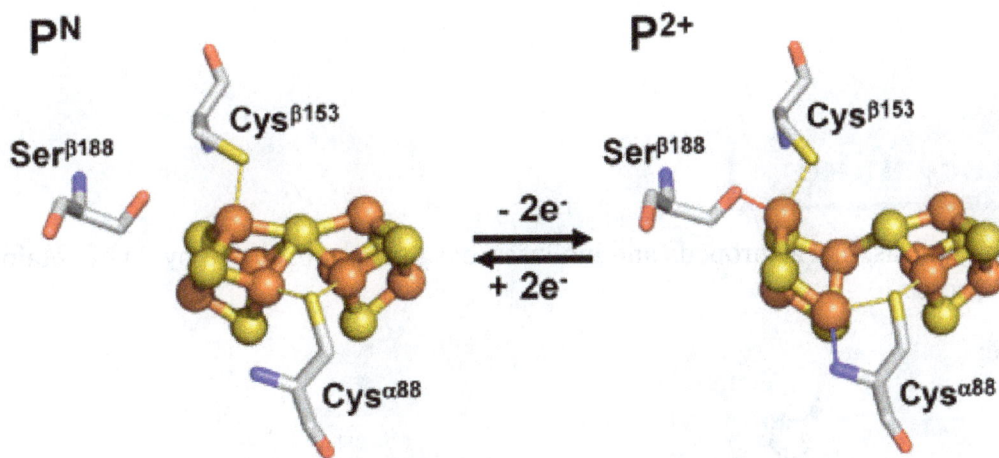

PN and P²⁺ clusters present in nitrogenase enzyme

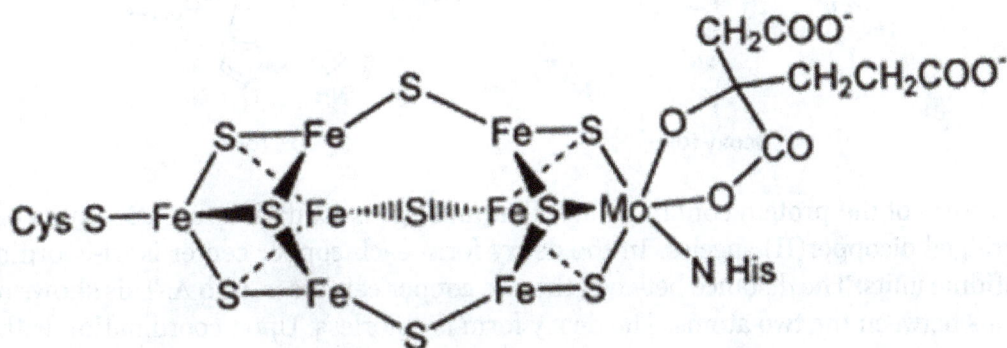

FeMo clusters present in nitrogenase enzyme

Proposed mechanism:

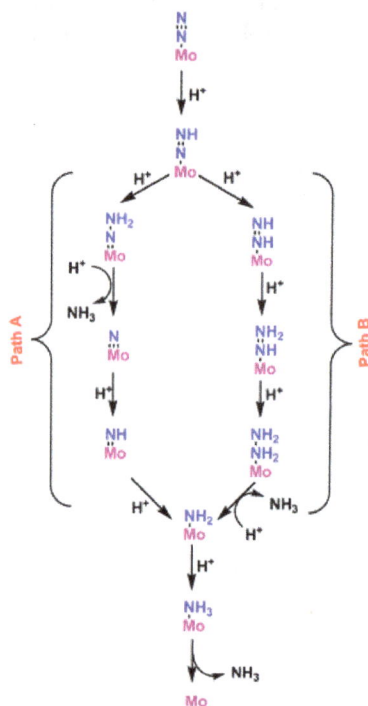

Hemocyanine

In many organisms, like arthropods and molluscs, oxygen is transported by Cu - containing hae-mocyanin protein.

Active site:

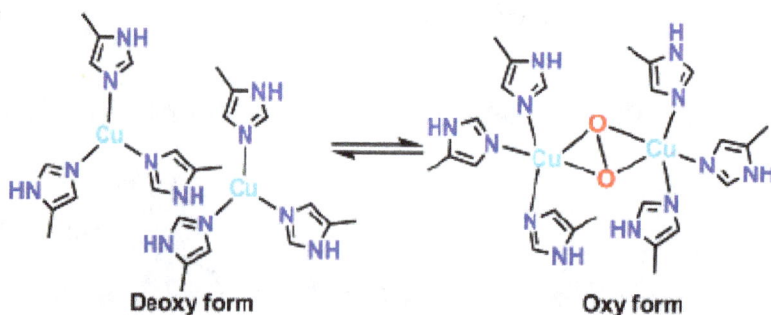

Deoxy form   Oxy form

The deoxy form of the protein contains Cu(I) that undergoes oxidation to Cu(II) in its oxy form, peroxo-bridged dicopper(II) species. In the deoxy form each copper center is tri-coordinated by three hintidine units. The distance between the two copper centers is 4.60 A. This shows no direct interactions between the two atoms. The deoxy form is colorless. Upon coordination with oxygen copper centers undergo to Cu(II) from Cu(I) and hence, a blue color appears due to oxo-to-Cu(II) charge transfer.

Tyrosinase:

Tyrosinase is a copper-containing protein. It catalyzes mono-oxygenation of phenol to diphenol or catechol.

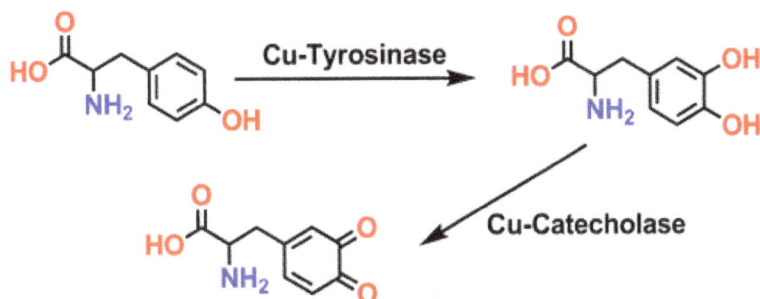

Active site:

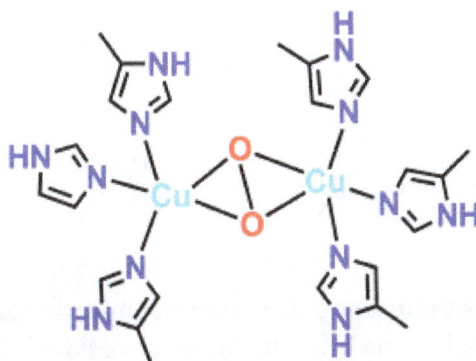

The deoxy form of this enzyme is a dinuclear Cu(I) where both copper centers are surrounded by three histidine residues. The two copper centers which are in +I oxidation state are about 2.4 A away from each other. In oxy form both copper centers oxidized to copper(II) oxidation state.

Mechanism

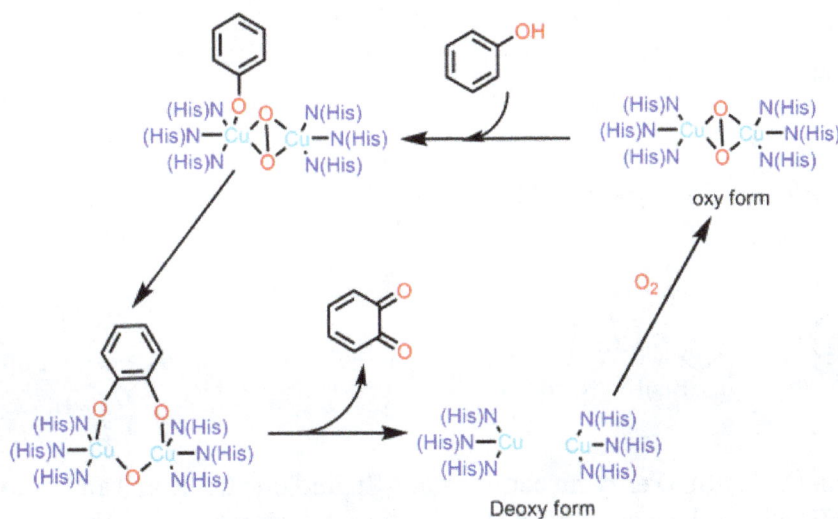

First the phenol part of tyrosine attached to one Cu(II) center after losing its proton. After that, an oxo insertion from Cu-oxo-Cu unit to the - ortho carbon atom to the phenol takes place. Oxidation of the 1,2-biphenol (catechol) to its quinine occurs with concomitant reduction of Cu(II) to Cu(I).

Cu-Zn Superoxide dismutase:

Superoxide dismutase catalyses the following dismutation reaction;

$$2O_2^- + 2H^+ \longrightarrow H_2O_2 + O_2$$

Active Site:

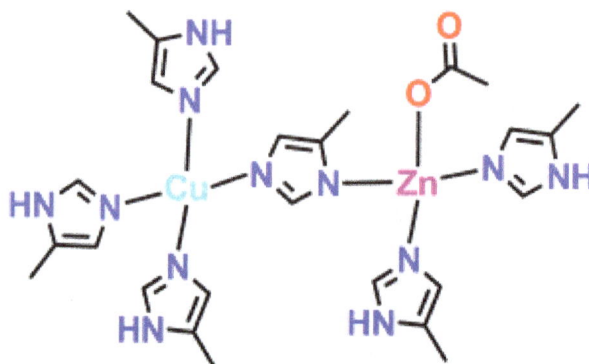

The active site of this enzyme is dinuclear and contains copper and zinc atom. The copper center is surrounded by three hitidine units, one bridging hinsitine residue and one axial weakly bound solvent (water) molecule. The zinc center is tetradentate and surrounded by two histidine residues, one one bridging hinsitine, and one Asp acid units. Copper and zinc centers are in +II oxidation state and about 6.0 A apart from each other.

Proposed mechanism:

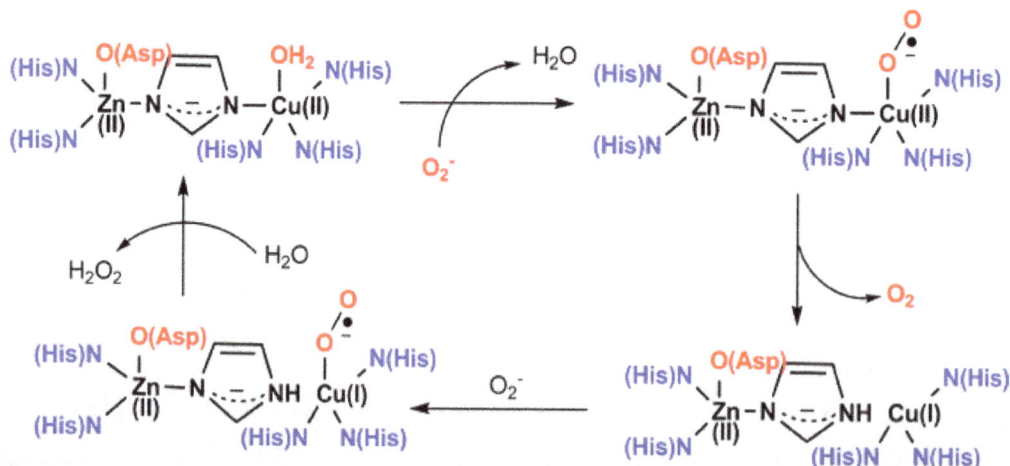

The active form of Cu/Zn-SOD enzyme contains one Zn and one Cu atoms and both are in +II oxidation state. Zn(II) center does not participates in the catalysis process, rather, its presence in the active site controls the reactivity shown by the Cu(II) center. One superoxide molecule, initially, binds to the Cu(II) centers and then hemolytic cleavage of the Cu-O bond reduces Cu(II) to Cu(I) and superoxide oxidizes to molecular oxygen. A vital structural change occurs upon transformation of Cu(II) to Cu(I). The Cu-N bond from the bridging histidine unit breaks. The second molecule of superoxide binds to Cu(I) and consequently, reduced to peroxide by Cu(I) that reoxidized to Cu(II). The labile peroxide is replaced by solvent water molecule.

Alcohol Dehydrogenase

Zn-containing mononuclear metalloenzyme alcohol dehydrogenase catalyzes interconversion of alcohol to aldehyde or ketone in the presence of NAD^+.

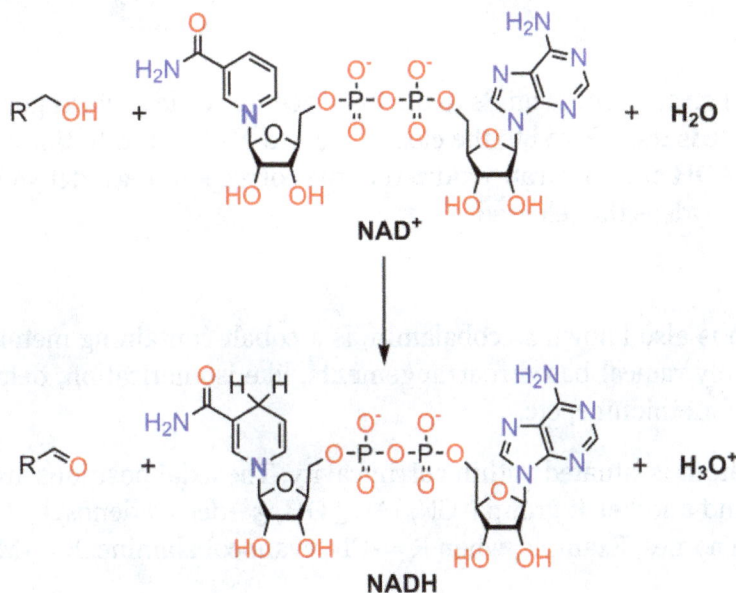

Active site:

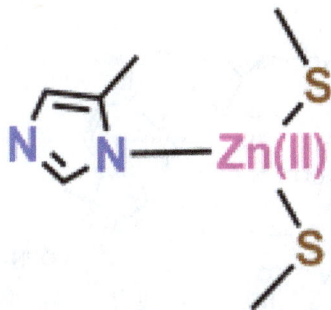

The active site of alcohol dehydrogenase contains Zn(II). The central Zn(II) center is surrounded by one nitrogen atom from a histidine imidazole residue and two sulfur atoms from two cystine moieties. The Zn(II) center acting as Lewis acid favors substrate to bind to the zinc center which then activate substrate.

Proposed mechanism:

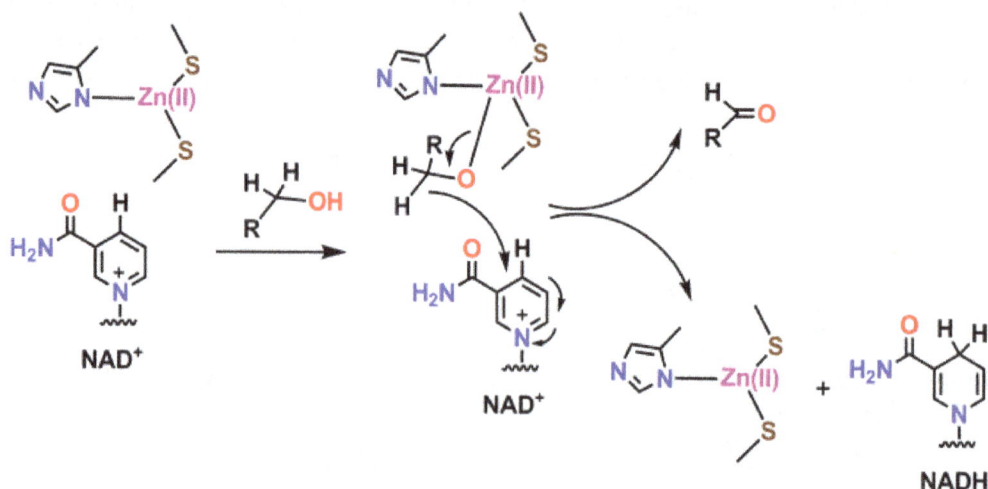

In the first phase substrate alcohol binds to the Zn(II) center which acts as good Lewis acid. In the second phase hydride is transferred to the closely located NAD+ unit. In this consequence process NAD+ reduced to NADH and substrate oxidized to the corresponding aldehyde. The weakly Zn(II) centered bound aldehyde is the released.

Vitamin B12:

Vitamin B12, which is also known as cobalamin, is a cobalt containing metalloenzyme. This enzyme catalyzes mainly radical-based rearrangements, like isomerization, dehydration, deamination, biosynthesis of methionine, etc.

The central cobalt atom is situated within corrin cavity. The axial positions are occupied by a imidazole derivative and another R group (-CN, Me, _OH, 5`-deoxyadenosyl, etc). Depending on R cobalamin has been names, Example, when R = -CN, cyanocobalamine, R = -Me, methylcobalami, etc.

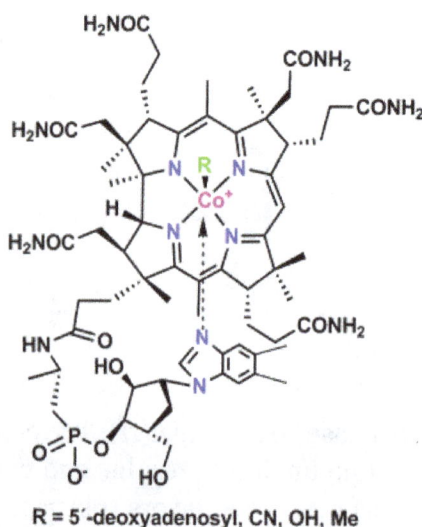

R = 5′-deoxyadenosyl, CN, OH, Me

Rearrangement reaction:

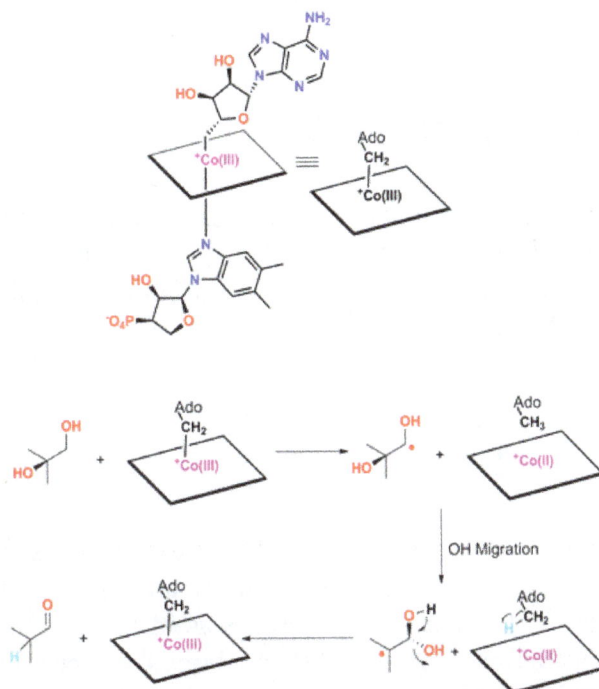

In the rearrangement reaction, in the first phase there is a hemolytic cleavage of Co-C bond in the cobalamin part. The formed CH_2-radical abstract hydrogen atom from the a -carbon atom and generates a carbon-radical center. After that, OH-radical migration takes place. The newly formed radical center abstract a hydrogen from the Ado-CH_3 center and further generates Ado-CH_2 radical that rebinds to the cobalt center. Loss of water from the substrate provides aldehyde product.

References

- Lee K, Jung J, Kim M, Guidotti G (January 2001). "Interaction of the alpha subunit of Na,K-ATPase with cofilin". The Biochemical Journal. 353 (2): 377–85. PMC 1221581. PMID 11139403. doi:10.1042/0264-6021:3530377

- Skou JC (February 1957). "The influence of some cations on an adenosine triphosphatase from peripheral nerves". Biochimica et Biophysica Acta. 23 (2): 394–401. PMID 13412736. doi:10.1016/0006-3002(57)90343-8

- Hall, John E.; Guyton, Arthur C. (2006). Textbook of medical physiology. St. Louis, Mo: Elsevier Saunders. ISBN 0-7216-0240-1

- Li Z, Cai T, Tian J, et al. (July 2009). "NaKtide, a Na/K-ATPase-derived peptide Src inhibitor, antagonizes ouabain-activated signal transduction in cultured cells". The Journal of Biological Chemistry. 284 (31): 21066–76

- Clausen MV, Hilbers F, Poulsen H (June 2017). "The Structure and Function of the Na,K-ATPase Isoforms in Health and Disease.". Frontiers in Physiology. 8. PMC 5459889. PMID 28634454. doi:10.3389/fphys.2017.00371

- Forrest, Michael (April 2015). "the_neuroscience_reason_we_fall_over_when_drunk". Science 2.0. Retrieved May 2015

- Howarth C, Gleeson P, Attwell D (July 2012). "Updated energy budgets for neural computation in the neocortex and cerebellum". J. Cereb. Blood Flow Metab. 32 (7): 1222–32. PMID 22434069. doi:10.1038/jcbfm.2012.35

Chemical Bonding in Inorganic Chemistry

Chemical bonds are the forces that bind atoms together to form a chemical compound. Chemical bonding is caused through electrostatic force or through the sharing of electrons as seen in co-valent bonds. The chapter closely examines the key concepts of chemical bonding to provide an extensive understanding of the subject.

Chemical Bonding

A chemical bond is a lasting attraction between atoms that enables the formation of chemical compounds. The bond may result from the electrostatic force of attraction between atoms with opposite charges, or through the sharing of electrons as in the covalent bonds. The strength of chemical bonds varies considerably; there are "strong bonds" or "primary bond" such as metallic, covalent or ionic bonds and "weak bonds" or "secondary bond" such as Dipole-dipole interaction, the London dispersion force and hydrogen bonding.

Since opposite charges attract via a simple electromagnetic force, the negatively charged electrons that are orbiting the nucleus and the positively charged protons in the nucleus attract each other. An electron positioned between two nuclei will be attracted to both of them, and the nuclei will be attracted toward electrons in this position. This attraction constitutes the chemical bond. Due to the matter wave nature of electrons and their smaller mass, they must occupy a much larger amount of volume compared with the nuclei, and this volume occupied by the electrons keeps the atomic nuclei in a bond relatively far apart, as compared with the size of the nuclei themselves.

In general, strong chemical bonding is associated with the sharing or transfer of electrons between the participating atoms. The atoms in molecules, crystals, metals and diatomic gases—indeed most of the physical environment around us—are held together by chemical bonds, which dictate the structure and the bulk properties of matter.

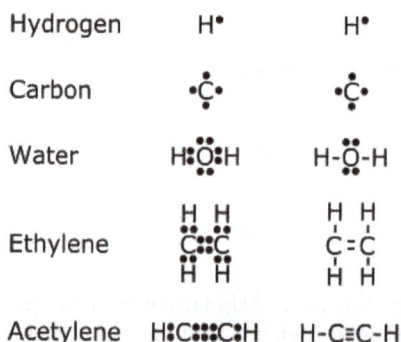

Hydrogen	H•	H•
Carbon	•C̈•	•C̈•
Water	H:Ö:H	H-Ö-H
Ethylene	H H C::C H H	H H C=C H H
Acetylene	H:C:::C:H	H-C≡C-H

Examples of Lewis dot-style representations of chemical bonds between carbon (C), hydrogen (H), and oxygen (O). Lewis dot diagrams were an early attempt to describe chemical bonding and are still widely used today.

All bonds can be explained by quantum theory, but, in practice, simplification rules allow chemists to predict the strength, directionality, and polarity of bonds. The octet rule and VSEPR theory are two examples. More sophisticated theories are valence bond theory which includes orbital hybridization and resonance, and molecular orbital theory which includes linear combination of atomic orbitals and ligand field theory. Electrostatics are used to describe bond polarities and the effects they have on chemical substances.

Overview of Main Types Of Chemical Bonds

A chemical bond is an attraction between atoms. This attraction may be seen as the result of different behaviors of the outermost or valence electrons of atoms. These behaviors merge into each other seamlessly in various circumstances, so that there is no clear line to be drawn between them. However it remains useful and customary to differentiate between different types of bond, which result in different properties of condensed matter.

In the simplest view of a covalent bond, one or more electrons (often a pair of electrons) are drawn into the space between the two atomic nuclei. Energy is released by bond formation. This is not as a reduction in potential energy, because the attraction of the two electrons to the two protons is offset by the electron-electron and proton-proton repulsions. Instead, the release of energy (and hence stability of the bond) arises from the reduction in kinetic energy due to the electrons being in a more spatially distributed (i.e. longer de Broglie wavelength) orbital compared with each electron being confined closer to its respective nucleus. These bonds exist between two particular identifiable atoms and have a direction in space, allowing them to be shown as single connecting lines between atoms in drawings, or modeled as sticks between spheres in models.

In a polar covalent bond, one or more electrons are unequally shared between two nuclei. Covalent bonds often result in the formation of small collections of better-connected atoms called molecules, which in solids and liquids are bound to other molecules by forces that are often much weaker than the covalent bonds that hold the molecules internally together. Such weak intermolecular bonds give organic molecular substances, such as waxes and oils, their soft bulk character, and their low melting points (in liquids, molecules must cease most structured or oriented contact with each other). When covalent bonds link long chains of atoms in large molecules, however (as in polymers such as nylon), or when covalent bonds extend in networks through solids that are not composed of discrete molecules (such as diamond or quartz or the silicate minerals in many types of rock) then the structures that result may be both strong and tough, at least in the direction oriented correctly with networks of covalent bonds. Also, the melting points of such covalent polymers and networks increase greatly.

In a simplified view of an *ionic* bond, the bonding electron is not shared at all, but transferred. In this type of bond, the outer atomic orbital of one atom has a vacancy which allows the addition of one or more electrons. These newly added electrons potentially occupy a lower energy-state (effectively closer to more nuclear charge) than they experience in a different atom. Thus, one nucleus offers a more tightly bound position to an electron than does another nucleus, with the result that one atom may transfer an electron to the other. This transfer causes one atom to assume a net positive charge, and the other to assume a net negative charge. The *bond* then results from electrostatic attraction between atoms and the atoms become positive or negatively charged ions. Ionic bonds may be seen as extreme examples of polarization in covalent bonds. Often, such bonds have no particular orientation in space, since they result from equal electrostatic attraction of each ion to all ions around them. Ionic bonds are strong (and thus ionic substances require high temperatures to melt) but also brittle, since the forces between ions are short-range and do not easily bridge cracks and fractures. This type of bond gives rise to the physical characteristics of crystals of classic mineral salts, such as table salt.

A less often mentioned type of bonding is *metallic* bonding. In this type of bonding, each atom in a metal donates one or more electrons to a "sea" of electrons that reside between many metal atoms. In this sea, each electron is free (by virtue of its wave nature) to be associated with a great many atoms at once. The bond results because the metal atoms become somewhat positively charged due to loss of their electrons while the electrons remain attracted to many atoms, without being part of any given atom. Metallic bonding may be seen as an extreme example of delocalization of electrons over a large system of covalent bonds, in which every atom participates. This type of bonding is often very strong (resulting in the tensile strength of metals). However, metallic bonding is more collective in nature than other types, and so they allow metal crystals to more easily deform, because they are composed of atoms attracted to each other, but not in any particularly-oriented ways. This results in the malleability of metals. The sea of electrons in metallic bonding causes the characteristically good electrical and thermal conductivity of metals, and also their "shiny" reflection of most frequencies of white light.

History

Early speculations about the nature of the chemical bond, from as early as the 12th century, supposed that certain types of chemical species were joined by a type of chemical affinity. In 1704, Sir Isaac Newton famously outlined his atomic bonding theory, in "Query 31" of his *Opticks*, whereby atoms attach to each other by some "force". Specifically, after acknowledging the various popular theories in vogue at the time, of how atoms were reasoned to attach to each other, i.e. "hooked atoms", "glued together by rest", or "stuck together by conspiring motions", Newton states that he would rather infer from their cohesion, that "particles attract one another by some force, which in immediate contact is exceedingly strong, at small distances performs the chemical operations, and reaches not far from the particles with any sensible effect."

In 1819, on the heels of the invention of the voltaic pile, Jöns Jakob Berzelius developed a theory of chemical combination stressing the electronegative and electropositive characters of the combining atoms. By the mid 19th century, Edward Frankland, F.A. Kekulé, A.S. Couper, Alexander Butlerov, and Hermann Kolbe, building on the theory of radicals, developed the theory of valency, originally called "combining power", in which compounds were joined owing to an attraction of positive and negative poles. In 1916, chemist Gilbert N. Lewis developed the concept of the electron-pair bond, in which two atoms may share one to six electrons, thus forming the single electron bond, a single bond, a double bond, or a triple bond; in Lewis's own words, "An electron may form a part of the shell of two different atoms and cannot be said to belong to either one exclusively."

That same year, Walther Kossel put forward a theory similar to Lewis' only his model assumed complete transfers of electrons between atoms, and was thus a model of ionic bonding. Both Lewis and Kossel structured their bonding models on that of Abegg's rule (1904).

In 1927, the first mathematically complete quantum description of a simple chemical bond, i.e.

that produced by one electron in the hydrogen molecular ion, H_2^+ was derived by the Danish physicist Oyvind Burrau. This work showed that the quantum approach to chemical bonds could be fundamentally and quantitatively correct, but the mathematical methods used could not be extended to molecules containing more than one electron. A more practical, albeit less quantitative, approach was put forward in the same year by Walter Heitler and Fritz London. The Heitler-London method forms the basis of what is now called valence bond theory. In 1929, the linear combination of atomic orbitals molecular orbital method (LCAO) approximation was introduced by Sir John Lennard-Jones, who also suggested methods to derive electronic structures of molecules of F_2 (fluorine) and O_2 (oxygen) molecules, from basic quantum principles. This molecular orbital theory represented a covalent bond as an orbital formed by combining the quantum mechanical Schrödinger atomic orbitals which had been hypothesized for electrons in single atoms. The equations for bonding electrons in multi-electron atoms could not be solved to mathematical perfection (i.e., *analytically*), but approximations for them still gave many good qualitative predictions and results. Most quantitative calculations in modern quantum chemistry use either valence bond or molecular orbital theory as a starting point, although a third approach, density functional theory, has become increasingly popular in recent years.

In 1933, H. H. James and A. S. Coolidge carried out a calculation on the dihydrogen molecule that, unlike all previous calculation which used functions only of the distance of the electron from the atomic nucleus, used functions which also explicitly added the distance between the two electrons. With up to 13 adjustable parameters they obtained a result very close to the experimental result for the dissociation energy. Later extensions have used up to 54 parameters and gave excellent agreement with experiments. This calculation convinced the scientific community that quantum theory could give agreement with experiment. However this approach has none of the physical pictures of the valence bond and molecular orbital theories and is difficult to extend to larger molecules.

Bonds in Chemical Formulas

Because atoms and molecules are three-dimensional, it is difficult to use a single method to indicate orbitals and bonds. In molecular formulas the chemical bonds (binding orbitals) between atoms are indicated in different ways depending on the type of discussion. Sometimes, some details are neglected. For example, in organic chemistry one is sometimes concerned only with the functional group of the molecule. Thus, the molecular formula of ethanol may be written in conformational form, three-dimensional form, full two-dimensional form (indicating every bond with no three-dimensional directions), compressed two-dimensional form (CH_3-CH_2-OH), by separating the functional group from another part of the molecule (C_2H_5OH), or by its atomic constituents (C_2H_6O), according to what is discussed. Sometimes, even the non-bonding valence shell electrons (with the two-dimensional approximate directions) are marked, e.g. for elemental carbon ˙C˙. Some chemists may also mark the respective orbitals, e.g. the hypothetical ethene⁻⁴ anion ($_\backslash/C=C_/\backslash$ ⁻⁴) indicating the possibility of bond formation.

Strong Chemical Bonds

Strong chemical bonds are the *intramolecular* forces which hold atoms together in molecules. A strong chemical bond is formed from the transfer or sharing of electrons between atomic centers and relies on the electrostatic attraction between the protons in nuclei and the electrons in the orbitals.

The types of strong bond differ due to the difference in electronegativity of the constituent elements. A large difference in electronegativity leads to more polar (ionic) character in the bond.

Ionic Bonding

Ionic bonding is a type of electrostatic interaction between atoms which have a large electronegativity difference. There is no precise value that distinguishes ionic from covalent bonding, but a difference of electronegativity of over 1.7 is likely to be ionic, and a difference of less than 1.7 is likely to be covalent. Ionic bonding leads to separate positive and negative ions. Ionic charges are commonly between −3e to +3e. Ionic bonding commonly occurs in metal salts such as sodium chloride (table salt). A typical feature of ionic bonds is that the species form into ionic crystals, in which no ion is specifically paired with any single other ion, in a specific directional bond. Rather, each species of ion is surrounded by ions of the opposite charge, and the spacing between it and each of the oppositely charged ions near it, is the same for all surrounding atoms of the same type. It is thus no longer possible to associate an ion with any specific other single ionized atom near it. This is a situation unlike that in covalent crystals, where covalent bonds between specific atoms are still discernible from the shorter distances between them, as measured via such techniques as X-ray diffraction.

Ionic crystals may contain a mixture of covalent and ionic species, as for example salts of complex acids, such as sodium cyanide, NaCN. X-ray diffraction shows that in NaCN, for example, the bonds between sodium cations (Na^+) and the cyanide anions (CN^-) are *ionic*, with no sodium ion associated with any particular cyanide. However, the bonds between C and N atoms in cyanide are of the *covalent* type, making each of the carbon and nitrogen associated with *just one* of its opposite type, to which it is physically much closer than it is to other carbons or nitrogens in a sodium cyanide crystal.

When such crystals are melted into liquids, the ionic bonds are broken first because they are non-directional and allow the charged species to move freely. Similarly, when such salts dissolve into water, the ionic bonds are typically broken by the interaction with water, but the covalent bonds continue to hold. For example, in solution, the cyanide ions, still bound together as single CN^- ions, move independently through the solution, as do sodium ions, as Na^+. In water, charged ions move apart because each of them are more strongly attracted to a number of water molecules, than to each other. The attraction between ions and water molecules in such solutions is due to a type of weak dipole-dipole type chemical bond. In melted ionic compounds, the ions continue to be attracted to each other, but not in any ordered or crystalline way.

Covalent Bond

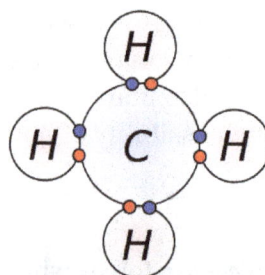

● Electron from hydrogen
● Electron from carbon

Nonpolar covalent bonds in methane (CH_4). The Lewis structure shows electrons shared between C and H atoms.

Covalent bonding is a common type of bonding, in which two atoms share two valence electrons, one from each of the atoms. In nonpolar covalent bonds, the electronegativity difference between the bonded atoms is small, typically 0 to 0.3. Bonds within most organic compounds are described as covalent. The figure shows methane (CH_4), in which each hydrogen forms a covalent bond with the carbon.

Molecules which are formed primarily from non-polar covalent bonds are often immiscible in water or other polar solvents, but much more soluble in non-polar solvents such as hexane.

A polar covalent bond is a covalent bond with a significant ionic character. This means that the two shared electrons are closer to one of the atoms than the other, creating an imbalance of charge. Such bonds occur between two atoms with moderately different electronegativities and give rise to dipole-dipole interactions. The electronegativity difference between the two atoms in these bonds is 0.3 to 1.7.

Single and Multiple Bonds

A single bond between two atoms corresponds to the sharing of one pair of electrons. The electron density of these two bonding electrons is concentrated in the region between the two atoms, which is the defining quality of a sigma bond.

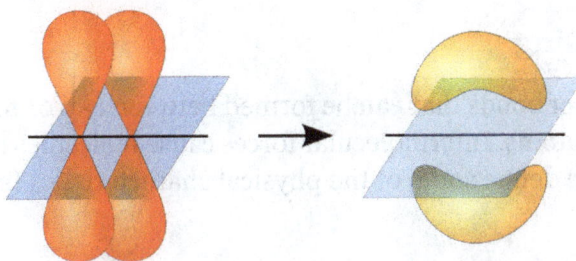

Two p-orbitals forming a pi-bond.

A double bond between two atoms is formed by the sharing of two pairs of electrons, one in a sigma bond and one in a pi bond, with electron density concentrated on two opposite sides of the internuclear axis. A triple bond consists of three shared electron pairs, forming one sigma and two pi bonds.

Quadruple and higher bonds are very rare and occur only between certain transition metal atoms.

Coordinate Covalent Bond (Dipolar Bond)

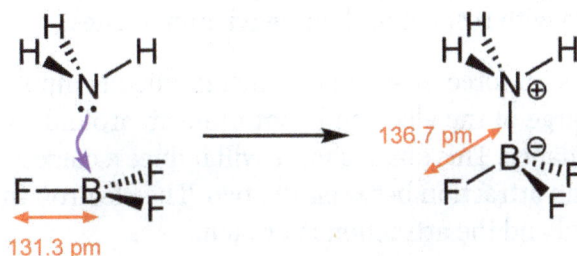

Adduct of ammonia and boron trifluoride

A coordinate covalent bond is a covalent bond in which the two shared bonding electrons are from the same one of the atoms involved in the bond. For example, boron trifluoride (BF_3) and ammonia (NH_3) from an adduct or coordination complex $F_3B \leftarrow NH_3$ with a B−N bond in which a lone pair of electrons on N is shared with an empty atomic orbital on B. BF_3 with an empty orbital is described as an electron pair acceptor or Lewis acid, while NH_3 with a lone pair which can be shared is described as an electron-pair donor or Lewis base. The electrons are shared roughly equally between the atoms in contrast to ionic bonding. Such bonding is shown by an arrow pointing to the Lewis acid.

Transition metal complexes are generally bound by coordinate covalent bonds. For example, the ion Ag^+ reacts as a Lewis acid with two molecules of the Lewis base NH_3 to form the complex ion $Ag(NH_3)_2^+$, which has two Ag←N coordinate covalent bonds.

Metallic Bonding

In metallic bonding, bonding electrons are delocalized over a lattice of atoms. By contrast, in ionic compounds, the locations of the binding electrons and their charges are static. The freely-moving or delocalization of bonding electrons leads to classical metallic properties such as luster (surface light reflectivity), electrical and thermal conductivity, ductility, and high tensile strength.

Intermolecular Bonding

There are four basic types of bonds that can be formed between two or more (otherwise non-associated) molecules, ions or atoms. Intermolecular forces cause molecules to be attracted or repulsed by each other. Often, these define some of the physical characteristics (such as the melting point) of a substance.

- A large difference in electronegativity between two bonded atoms will cause a permanent charge separation, or dipole, in a molecule or ion. Two or more molecules or ions with permanent dipoles can interact within dipole-dipole interactions. The bonding electrons in a molecule or ion will, on average, be closer to the more electronegative atom more frequently than the less electronegative one, giving rise to partial charges on each atom, and causing electrostatic forces between molecules or ions.

- A hydrogen bond is effectively a strong example of an interaction between two permanent dipoles. The large difference in electronegativities between hydrogen and any of fluorine, nitrogen and oxygen, coupled with their lone pairs of electrons cause strong electrostatic forces between molecules. Hydrogen bonds are responsible for the high boiling points of water and ammonia with respect to their heavier analogues.

- The London dispersion force arises due to instantaneous dipoles in neighbouring atoms. As the negative charge of the electron is not uniform around the whole atom, there is always a charge imbalance. This small charge will induce a corresponding dipole in a nearby molecule; causing an attraction between the two. The electron then moves to another part of the electron cloud and the attraction is broken.

- A cation–pi interaction occurs between a pi bond and a cation.

Theories of Chemical Bonding

In the (unrealistic) limit of "pure" ionic bonding, electrons are perfectly localized on one of the two atoms in the bond. Such bonds can be understood by classical physics. The forces between the atoms are characterized by isotropic continuum electrostatic potentials. Their magnitude is in simple proportion to the charge difference.

Covalent bonds are better understood by valence bond theory or molecular orbital theory. The properties of the atoms involved can be understood using concepts such as oxidation number. The electron density within a bond is not assigned to individual atoms, but is instead delocalized between atoms. In valence bond theory, the two electrons on the two atoms are coupled together with the bond strength depending on the overlap between them. In molecular orbital theory, the linear combination of atomic orbitals (LCAO) helps describe the delocalized molecular orbital structures and energies based on the atomic orbitals of the atoms they came from. Unlike pure ionic bonds, covalent bonds may have directed anisotropic properties. These may have their own names, such as sigma bond and pi bond.

In the general case, atoms form bonds that are intermediate between ionic and covalent, depending on the relative electronegativity of the atoms involved. This type of bond is sometimes called polar covalent.

Bonding

The stability of the monoatomic noble gases emphasized the fact that an atomic system with eight outer most electrons (two for He) will be very stable and called as noble gas configuration. This is also known as octet rule. The above statement can be further described for the inclusion of He as, the atomic system with filled outermost shell will be stable.

Atoms are connected to another homo - or hetero - atom(s) forming molecules, while, molecules of noble gases are monoatomic. This indicates that a molecule will only be formed by the combination of either homo - or hetero - atom if it tends to a low energy, filled-configuration of the outermost shell, and more stable systems than individual atomic systems.

In a di or polyatomic molecules, atoms are held together by means of an attraction force. This force is called bond. A bond could be formed by equal sharing of electron density between atoms (covalent bonding), or uneven sharing of electron density together with coulomb interaction between the atoms (ionic bonding). Whatever, a stable electronic configuration should be achieved by atom either by releasing/accepting electron(s) or sharing electron(s).

The inter-nuclear distance between two atoms is known as bond distance. The bond dissociation energy is the enthalpy change for complete splitting the units linked by a particular chemical bond in the gaseous state. It is also defined as the amount of energy released when the bond is formed between two neutral gaseous atoms.

Types of bonds:

Elements can be subdivided mainly into two groups; electropositive and electronegative elements.

Those elements which release or accept electron(s) to achieve an inert gas configuration are called electropositive and electronegative elements, respectively. Combination of electropositive and electronegative elements forms the following compounds.

(A) Ionic compounds = Ionic bond = Electropositive elements + Electronegative elements

(B) Covalent Compounds = Covalent bond = Electronegative elements + Electronegative elements

(C) Metallic Compounds = Metallic bond = Electropositive elements + Electropositive element

Examples:

Covalent Homoatomic bonding

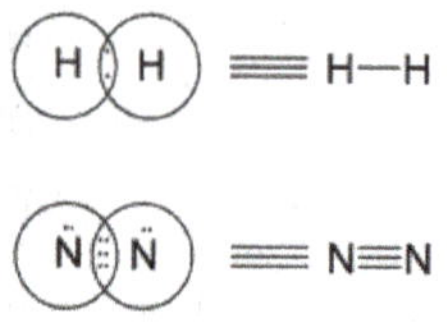

Partial Covalent Heteroatomic bonding

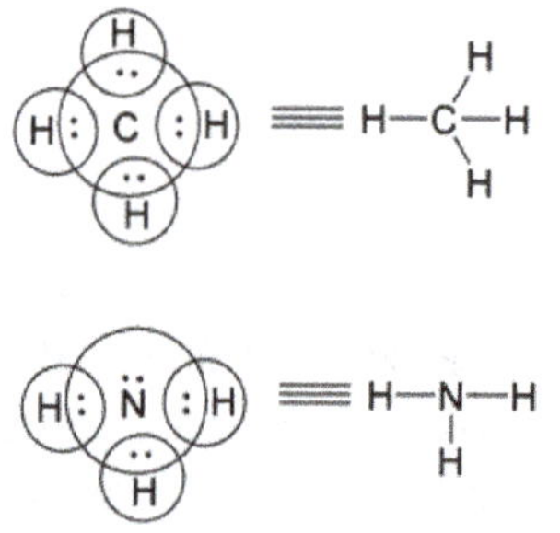

Ionic bonding

$$Na(1s^2\,2s^2\,2p^6\,3s^1) \xrightarrow{-1e-} Na^+(1s^2\,2s^2\,2p^6)$$

n=2 with 8 electrons
i.e. stable configuration
noble gas configuration $\Big\}$ NaCl

$$Cl(1s^2\,2s^2\,2p^6\,3s^2\,3p^5) \xrightarrow{+1e-} Cl^+(1s^2\,2s^2\,2p^6\,3s^2\,3p^6)$$

n=3 with 8 electrons
i.e. stable configuration,
noble gas configuration

Covalent, partial covalent and ionic bonding patterns maintaining octet rule.

Ionic Bonding

Ionic bonding is a type of chemical bond that involves the electrostatic attraction between oppositely charged ions, and is the primary interaction occurring in ionic compounds. The ions are atoms that have gained one or more electrons (known as anions, which are negatively charged) and atoms that have lost one or more electrons (known as cations, which are positively charged). This transfer of electrons is known as electrovalence in contrast to covalence. In the simplest case, the cation is a metal atom and the anion is a nonmetal atom, but these ions can be of a more complex nature, e.g. molecular ions like NH_4^+ or SO_4^{2-}. In simpler words, an ionic bond is the transfer of electrons from a metal to a non-metal in order to obtain a full valence shell for both atoms.

It is important to recognize that *clean* ionic bonding – in which one atom or molecule completely share an electron from another – cannot exist: all ionic compounds have some degree of covalent bonding, or electron sharing. Thus, the term "ionic bonding" is given when the ionic character is greater than the covalent character – that is, a bond in which a large electronegativity difference exists between the two atoms, causing the bonding to be more polar (ionic) than in covalent bonding where electrons are shared more equally. Bonds with partially ionic and partially covalent character are called polar covalent bonds.

Ionic compounds conduct electricity when molten or in solution, typically as a solid. Ionic compounds generally have a high melting point, depending on the charge of the ions they consist of. The higher the charges the stronger the cohesive forces and the higher the melting point. They also tend to be soluble in water. Here, the opposite trend roughly holds: the weaker the cohesive forces, the greater the solubility.

Overview

Atoms that have an almost full or almost empty valence shell tend to be very reactive. Atoms that are strongly electronegative (as is the case with halogens) often have only one or two empty orbitals in their valence shell, and frequently bond with other molecules or gain electrons to form anions. Atoms that are weakly electronegative (such as alkali metals) have relatively few valence electrons that can easily be shared with atoms that are strongly electronegative. As a result, weakly electronegative atoms tend to distort their electrons cloud and form cations.

Formation

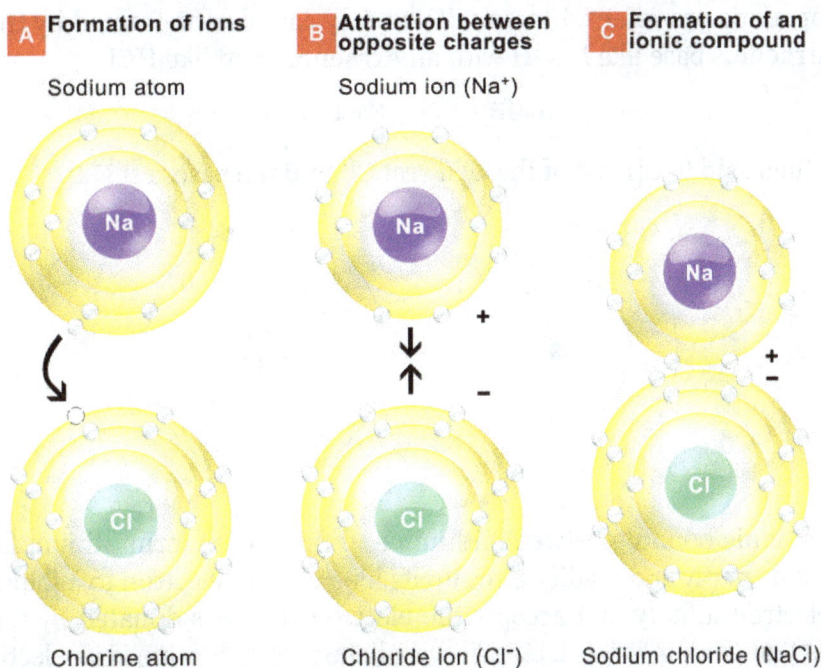

Formation of an Ionic Bond

Ionic bonds in sodium chloride

Ionic bonding can result from a redox reaction when atoms of an element (usually metal), whose ionization energy is low, give some of their electrons to achieve a stable electron configuration. In doing so, cations are formed. The atom of another element (usually nonmetal), whose electron affinity is positive, then accepts the electron(s), again to attain a stable electron configuration, and after accepting electron(s) the atom becomes an anion. Typically, the stable electron configuration is one of the noble gases for elements in the s-block and the p-block, and particular stable electron configurations for d-block and f-block elements. The electrostatic attraction between the anions and cations leads to the formation of a solid with a crystallographic lattice in which the ions are stacked in an alternating fashion. In such a lattice, it is usually not possible to distinguish discrete molecular units, so that the compounds formed are not molecular in nature. However, the ions themselves can be complex and form molecular ions like the acetate anion or the ammonium cation.

For example, common table salt is sodium chloride. When sodium (Na) and chlorine (Cl) are combined, the sodium atoms each lose an electron, forming cations (Na^+), and the chlorine atoms each gain an electron to form anions (Cl^-). These ions are then attracted to each other in a 1:1 ratio to form sodium chloride (NaCl).

$$Na + Cl \rightarrow Na^+ + Cl^- \rightarrow NaCl$$

However, to maintain charge neutrality, strict ratios between anions and cations are observed so that ionic compounds, in general, obey the rules of stoichiometry despite not being molecular compounds. For compounds that are transitional to the alloys and possess mixed ionic and metallic bonding, this may not be the case anymore. Many sulfides, e.g., do form non-stoichiometric compounds.

Many ionic compounds are referred to as salts as they can also be formed by the neutralization reaction of an Arrhenius base like NaOH with an Arrhenius acid like HCl

$$NaOH + HCl \rightarrow NaCl + H_2O$$

The salt NaCl is then said to consist of the acid rest Cl^- and the base rest Na^+.

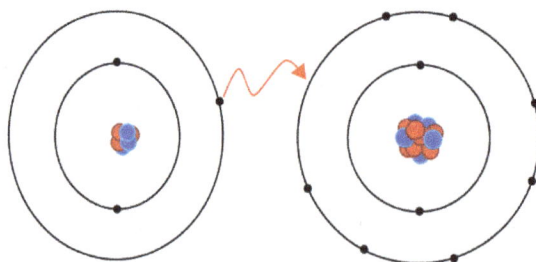

Representation of ionic bonding between lithium and fluorine to form lithium fluoride. Lithium has a low ionization energy and readily gives up its lone valence electron to a fluorine atom, which has a positive electron affinity and accepts the electron that was donated by the lithium atom. The end-result is that lithium is isoelectronic with helium and fluorine is isoelectronic with neon. Electrostatic interaction occurs between the two resulting ions, but typically aggregation is not limited to two of them. Instead, aggregation into a whole lattice held together by ionic bonding is the result.

The removal of electrons from the cation is endothermic, raising the system's overall energy. There may also be energy changes associated with breaking of existing bonds or the addition of more than one electron to form anions. However, the action of the anion's accepting the cation's valence electrons and the subsequent attraction of the ions to each other releases (lattice) energy and, thus, lowers the overall energy of the system.

Ionic bonding will occur only if the overall energy change for the reaction is favorable. In general, the reaction is exothermic, but, e.g., the formation of mercuric oxide (HgO) is endothermic. The charge of the resulting ions is a major factor in the strength of ionic bonding, e.g. a salt C^+A^- is held together by electrostatic forces roughly four times weaker than $C^{2+}A^{2-}$ according to Coulombs law, where C and A represent a generic cation and anion respectively. Of course the sizes of the ions and the particular packing of the lattice are ignored in this simple argument.

Structures

Ionic compounds in the solid state form lattice structures. The two principal factors in determining the form of the lattice are the relative charges of the ions and their relative sizes. Some structures are adopted by a number of compounds; for example, the structure of the rock salt sodium chloride is also adopted by many alkali halides, and binary oxides such as MgO. Pauling's rules provide guidelines for predicting and rationalizing the crystal structures of ionic crystals.

Bond Strength

For a solid crystalline ionic compound the enthalpy change in forming the solid from gaseous ions is termed the lattice energy. The experimental value for the lattice energy can be determined using the Born-Haber cycle. It can also be calculated (predicted) using the Born-Landé equation as the sum of the electrostatic potential energy, calculated by summing interactions between cations and anions, and a short-range repulsive potential energy term. The electrostatic potential can be expressed in terms of the inter-ionic separation and a constant (Madelung constant) that takes account of the geometry of the crystal. The further away from the nucleus the weaker the shield. The Born-Landé equation gives a reasonable fit to the lattice energy of, e.g., sodium chloride, where the calculated (predicted) value is -756 kJ/mol, which compares to -787 kJ/mol using the Born-Haber cycle.

Polarization Effects

Ions in crystal lattices of purely ionic compounds are spherical; however, if the positive ion is small and/or highly charged, it will distort the electron cloud of the negative ion, an effect summarised in Fajans' rules. This polarization of the negative ion leads to a build-up of extra charge density between the two nuclei, i.e., to partial covalency. Larger negative ions are more easily polarized, but the effect is usually important only when positive ions with charges of 3+ (e.g., Al^{3+}) are involved. However, 2+ ions (Be^{2+}) or even 1+ (Li^+) show some polarizing power because their sizes are so small (e.g., LiI is ionic but has some covalent bonding present). Note that this is not the ionic polarization effect that refers to displacement of ions in the lattice due to the application of an electric field.

Comparison with Covalent Bonding

In ionic bonding, the atoms are bound by attraction of opposite ions, whereas, in covalent bonding, atoms are bound by sharing electrons to attain stable electron configurations. In covalent bonding, the molecular geometry around each atom is determined by valence shell electron pair repulsion VSEPR rules, whereas, in ionic materials, the geometry follows maximum packing rules. One could say that covalent bonding is more *directional* in the sense that the energy penalty for not adhering to the optimum bond angles is large, whereas ionic bonding has no such penalty. There are no shared electron pairs to repel each other, the ions should simply be packed as efficiently as possible. This often leads to much higher coordination numbers. In NaCl, each ion has 6 bonds and all bond angles are 90 degrees. In CsCl the coordination number is 8. By comparison carbon typically has a maximum of four bonds.

Purely ionic bonding cannot exist, as the proximity of the entities involved in the bonding allows some degree of sharing electron density between them. Therefore, all ionic bonding has some covalent character. Thus, bonding is considered ionic where the ionic character is greater than the covalent character. The larger the difference in electronegativity between the two types of atoms involved in the bonding, the more ionic (polar) it is. Bonds with partially ionic and partially covalent character are called polar covalent bonds. For example, Na–Cl and Mg–O interactions have a few percent covalency, while Si–O bonds are usually ~50% ionic and ~50% covalent. Pauling estimated that an electronegativity difference of 1.7 (on the Pauling scale) corresponds to 50% ionic character, so that a difference greater than 50% corresponds to a bond which is predominantly ionic. Ionic character in covalent bonds can be directly measured for atoms having quadrupolar nuclei (^{2}H, ^{14}N, ^{81,79}Br, ^{35,37}Cl or ^{127}I). These nuclei are generally objects of NQR nuclear quadrupole resonance and NMR nuclear magnetic resonance studies. Interactions between the nuclear quadrupole moments Q and the electric field gradients (EFG) are characterized via the nuclear quadrupole coupling constants $QCC = e^2q_{zz}Q/h$ where the eq_{zz} term corresponds to the principal component of the EFG tensor and e is the elementary charge. In turn, the electric field gradient opens the way to description of bonding modes in molecules when the QCC values are accurately determined by NMR or NQR methods.

In general, when ionic bonding occurs in the solid (or liquid) state, it is not possible to talk about a single "ionic bond" between two individual atoms, because the cohesive forces that keep the lattice together are of a more collective nature. This is quite different in the case of covalent bonding, where we can often speak of a distinct bond localized between two particular atoms. However, even if ionic bonding is combined with some covalency, the result is *not* necessarily discrete bonds of a localized character. In such cases, the resulting bonding often requires description in terms of a band structure consisting of gigantic molecular orbitals spanning the entire crystal. Thus, the bonding in the solid often retains its collective rather than localized nature. When the difference in electronegativity is decreased, the bonding may then lead to a semiconductor, a semimetal or eventually a metallic conductor with metallic bonding.

Ionic Compound

In chemistry, an ionic compound is a chemical compound composed of ions held together by electrostatic forces termed ionic bonding. The compound is neutral overall, but consists of positively charged ions called cations and negatively charged ions called anions. These can be simple ions such as the sodium (Na^+) and chloride (Cl^-) in sodium chloride, or polyatomic species such as the ammonium (NH_4^+) and carbonate (CO_3^{2-}) ions in ammonium carbonate. Individual ions within an ionic compound usually have multiple nearest neighbours, so are not considered to be part of molecules, but instead part of a continuous three-dimensional network, usually in a crystalline structure.

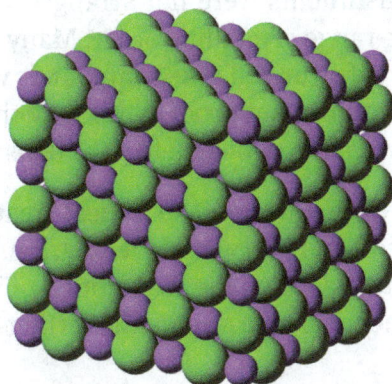

The crystal structure of sodium chloride, NaCl, a typical ionic compound. The purple spheres represent sodium cations, Na^+, and the green spheres represent chloride anions, Cl^-.

Ionic compounds containing hydrogen ions (H^+) are classified as acids, and those containing basic ions hydroxide (OH^-) or oxide (O^{2-}) are classified as bases. Ionic compounds without these ions are also known as salts and can be formed by acid–base reactions. Ionic compounds can also be produced from their constituent ions by evaporation of their solvent, precipitation, freezing, a solid-state reaction, or the electron transfer reaction of reactive metals with reactive non-metals, such as halogen gases.

Ionic compounds typically have high melting and boiling points, and are hard and brittle. As solids they are almost always electrically insulating, but when melted or dissolved they become highly conductive, because the ions are mobilized.

History of Discovery

This term was introduced by English physicist and chemist Michael Faraday in 1834 for the then-unknown species that *goes* from one electrode to the other through an aqueous medium.

X-ray spectrometer developed by Bragg

In 1913 the crystal structure of sodium chloride was determined by William Henry Bragg and William Lawrence Bragg. This revealed that there were six equidistant nearest-neighbours for each atom, demonstrating that the constituents were not arranged in molecules or finite aggregates, but instead as a network with long-range crystalline order. Many other inorganic compounds were also found to have similar structural features. These compounds were soon described as being constituted of ions rather than neutral atoms, but proof of this hypothesis was not found until the mid-1920s, when X-ray reflection experiments (which detect the density of electrons), were performed.

Principal contributors to the development of a theoretical treatment of ionic crystal structures were Max Born, Fritz Haber, Alfred Landé, Erwin Madelung, Paul Peter Ewald, and Kazimierz Fajans. Born predicted crystal energies based on the assumption of ionic constituents, which showed good correspondence to thermochemical measurements, further supporting the assumption.

Formation

Halite, the mineral form of sodium chloride, forms when salty water evaporates leaving the ions behind.

Ionic compounds can be produced from their constituent ions by evaporation, precipitation, or freezing. Reactive metals such as the alkali metals can react directly with the highly electronegative halogen gases to form an ionic product. They can also be synthesized as the product of a high temperature reaction between solids.

If the ionic compound is soluble in a solvent, it can be obtained as a solid compound by evaporating the solvent from this electrolyte solution. As the solvent is evaporated, the ions do not go into the vapour, but stay in the remaining solution, and when they become sufficiently concentrated, nucleation occurs, and they crystallize into an ionic compound. This process occurs widely in nature, and is the means of formation of the evaporite minerals. Another method of recovering the compound from solution involves saturating a solution at high temperature and then reducing the solubility by reducing the temperature until the solution is supersaturated and the solid compound nucleates.

Insoluble ionic compounds can be precipitated by mixing two solutions, one with the cation and one with the anion in it. Because all solutions are electrically neutral, the two solutions mixed must also contain counterions of the opposite charges. To ensure that these do not contaminate the precipitated ionic compound, it is important to ensure they do not also precipitate. If the two solutions have hydrogen ions and hydroxide ions as the counterions, they will react with one another in what is called an acid–base reaction or a neutralization reaction to form water. Alternately the counterions can be chosen to ensure that even when combined into a single solution they will remain soluble as spectator ions.

If the solvent is water in either the evaporation or precipitation method of formation, in many cases the ionic crystal formed also includes water of crystallization, so the product is known as a hydrate, and can have very different chemical properties.

Molten salts will solidify on cooling to below their freezing point. This is sometimes used for the solid-state synthesis of complex ionic compounds from solid reactants, which are first melted together. In other cases, the solid reactants do not need to be melted, but instead can react through a solid-state reaction route. In this method the reactants are repeatedly finely ground into a paste, and then heated to a temperature where the ions in neighbouring reactants can diffuse together during the time the reactant mixture remains in the oven. Other synthetic routes use a solid precursor with the correct stoichiometric ratio of non-volatile ions, which is heated to drive off other species.

In some reactions between highly reactive metals (usually from Group 1 or Group 2) and highly electronegative halogen gases, or water, the atoms can be ionized by electron transfer, a process thermodynamically understood using the Born–Haber cycle.

Bonding

Ions in ionic compounds are primarily held together by the electrostatic forces between the charge distribution of these bodies, and in particular the ionic bond resulting from the long-ranged Coulomb attraction between the net negative charge of the anions and net positive charge of the cations. There is also a small additional attractive force from van der Waals interactions which contributes only around 1–2% of the cohesive energy for small ions. When a pair of ions comes close enough for their outer electron shells (most simple ions have closed shells) to overlap, a short-ranged repulsive force occurs, due to the Pauli exclusion principle. The balance between these forces leads to a potential energy well with a minimum energy when the nuclei are separated by a specific equilibrium distance.

If the electronic structure of the two interacting bodies is affected by the presence of one another, covalent interactions (non-ionic) also contribute to the overall energy of the compound formed. Ionic compounds are rarely purely ionic, i.e. held together only by electrostatic forces. The bonds between even the most electronegative/electropositive pairs such as those in caesium fluoride exhibit a small degree of covalency. Conversely, covalent bonds between unlike atoms often exhibit some charge separation and can be considered to have a partial ionic character. The circumstances under which a compound will have ionic or covalent character can typically be understood using Fajans' rules, which use only charges and the sizes of each ion. According to these rules, compounds with the most ionic character will have large positive ions with a low charge, bonded to a small negative ion with a high charge. More generally HSAB theory can be applied, whereby the compounds with the most ionic character are those consisting of hard acids and hard bases: small, highly charged ions with a high difference in electronegativities between the anion and cation. This difference in electronegativities means that the charge separation, and resulting dipole moment, is maintained even when the ions are in contact (the excess electrons on the anions are not transferred or polarized to neutralize the cations).

Structure

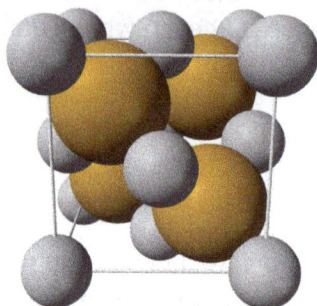

The unit cell of the zinc blende structure

Ions typically pack into extremely regular crystalline structures, in an arrangement that minimizes the lattice energy (maximizing attractions and minimizing repulsions). The lattice energy is the summation of the interaction of all sites with all other sites. For unpolarizable spherical ions only the charges and distances are required to determine the electrostatic interaction energy. For any particular ideal crystal structure, all distances are geometrically related to the smallest internuclear distance. So for each possible crystal structure, the total electrostatic energy can be related to the electrostatic energy of unit charges at the nearest neighbour distance by a multiplicative constant called the Madelung constant that can be efficiently computed using an Ewald sum. When a reasonable form is assumed for the additional repulsive energy, the total lattice energy can be modelled using the Born–Landé equation, the Born–Mayer equation, or in the absence of structural information, the Kapustinskii equation.

Using an even simpler approximation of the ions as impenetrable hard spheres, the arrangement of anions in these systems are often related to close-packed arrangements of spheres, with the cations occupying tetrahedral or octahedral interstices. Depending on the stoichiometry of the ionic compound, and the coordination (principally determined by the radius ratio) of cations and anions, a variety of structures are commonly observed, and theoretically rationalized by Pauling's rules.

Common ionic compound structures with close-packed anions							
Stoichiometry	Cation:anion coordination	Interstitial sites		Cubic close packing of anions		Hexagonal close packing of anions	
		occupancy	critical radius ratio	name	Madelung constant	name	Madelung constant
MX	6:6	all octahedral	0.4142	sodium chloride	1.747565	nickeline	<1.73[a]
	4:4	alternate tetrahedral	0.2247	zinc blende	1.6381	wurtzite	1.641
MX$_2$	8:4	all tetrahedral	0.2247	fluorite	5.03878		
	6:3	half octahedral (alternate layers fully occupied)	0.4142	cadmium chloride	5.61	cadmium iodide	4.71

MX_3	6:2	one-third octahedral	0.4142	rhodium(III) bromide[b]	6.67[c]	bismuth iodide	8.26[d]
M_2X_3	6:4	two-thirds octahedral	0.4142			corundum	25.0312
ABO_3		two-thirds octahedral	0.4142			ilmenite	depends on charges and structure [e]
AB_2O_4		one-eighth tetrahedral and one-half octahedral	$r_A/r_O = 0.2247$, $r_B/r_O = 0.4142$[f]	spinel, inverse spinel	depends on cation site distributions	olivine	depends on cation site distributions

In some cases the anions take on a simple cubic packing, and the resulting common structures observed are:

Common ionic compound structures with simple cubic packed anions					
Stoichiometry	Cation:anion coordination	Interstitial sites occupied	Example structure		
			name	critical radius ratio	Madelung constant
MX	8:8	entirely filled	cesium chloride	0.7321	1.762675
MX_2	8:4	half filled	calcium fluoride		
M_2X	4:8	half filled	lithium oxide		

Some ionic liquids, particularly with mixtures of anions or cations, can be cooled rapidly enough that there is not enough time for crystal nucleation to occur, so an ionic glass is formed (with no long-range order).

Defects

Frenkel defect   Schottky defect

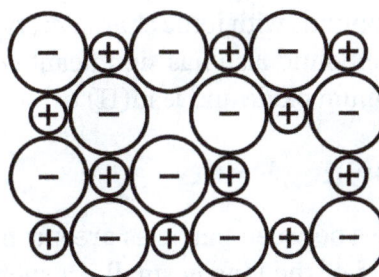

Within an ionic crystal, there will usually be some point defects, but to maintain electroneutrality, these defects come in pairs. Frenkel defects consist of a cation vacancy paired with a cation interstitial and can be generated anywhere in the bulk of the crystal, occurring most commonly in compounds with a low coordination number and cations that are much smaller than the anions. Schottky defects consist of one vacancy of each type, and are generated at the surfaces of a crystal, occurring most commonly in compounds with a high coordination number and when the anions and cations are of similar size. If the cations have multiple possible oxidation states, then it is possible for cation vacancies to compensate for electron deficiencies on cation sites with higher oxidation numbers, resulting in a non-stoichiometric compound. Another non-stoichiometric possibility is the formation of an F-center, a free electron occupying an anion vacancy. When the compound has three or more ionic components, even more defect types are possible. All of these point defects can be generated via thermal vibrations and have an equilibrium concentration. Because they are energetically costly, but entropically beneficial, they occur in greater concentration at higher temperatures. Once generated, these pairs of defects can diffuse mostly independently of one another, by hopping between lattice sites. This defect mobility is the source of most transport phenomena within an ionic crystal, including diffusion and solid state ionic conductivity. When vacancies collide with interstitials (Frenkel), they can recombine and annihilate one another. Similarly vacancies are removed when they reach the surface of the crystal (Schottky). Defects in the crystal structure generally expand the lattice parameters, reducing the overall density of the crystal. Defects also result in ions in distinctly different local environments, which causes them to experience a different crystal-field symmetry, especially in the case of different cations exchanging lattice sites. This results in a different splitting of d-electron orbitals, so that the optical absorption (and hence colour) can change with defect concentration.

Properties

Acidity/basicity

Ionic compounds containing hydrogen ions (H^+) are classified as acids, and those containing electropositive cations and basic anions ions hydroxide (OH^-) or oxide (O^{2-}) are classified as bases. Other ionic compounds are known as salts and can be formed by acid–base reactions. If the compound is the result of a reaction between a strong acid and a weak base, the result is an acidic salt. If it is the result of a reaction between a strong base and a weak acid, the result is a basic salt. If it is the result of a reaction between a strong acid and a strong base, the result is a neutral salt. Weak acids reacted with weak bases can produce ionic compounds with both the conjugate base ion and conjugate acid ion, such as ammonium acetate.

Some ions are classed as amphoteric, being able to react with either an acid or a base. This is also true of some compounds with ionic character, typically oxides or hydroxides of less-electropositive metals (so the compound also has significant covalent character), such as zinc oxide, aluminium hydroxide, aluminium oxide and lead(II) oxide.

Melting and Boiling Points

Electrostatic forces between particles are strongest when the charges are high, and the distance between the nuclei of the ions is small. In such cases, the compounds generally have very high melting and boiling points and a low vapour pressure. Trends in melting points can be even better explained when the structure and ionic size ratio is taken into account. Above their melting point ionic solids melt and become molten salts (although some ionic compounds such as aluminium chloride and iron(III) chloride show molecule-like structures in the liquid phase). Inorganic compounds with simple ions typically have small ions, and thus have high melting points, so are solids at room temperature. Some substances with larger ions, however, have a melting point below or near room temperature (often defined as up to 100 °C), and are termed ionic liquids. Ions in ionic liquids often have uneven charge distributions, or bulky substituents like hydrocarbon chains, which also play a role in determining the strength of the interactions and propensity to melt.

Even when the local structure and bonding of an ionic solid is disrupted sufficiently to melt it, there are still strong long-range electrostatic forces of attraction holding the liquid together and preventing ions boiling to form a gas phase. This means that even room temperature ionic liquids have low vapour pressures, and require substantially higher temperatures to boil. Boiling points exhibit similar trends to melting points in terms of the size of ions and strength of other interactions. When vapourized, the ions are still not freed of one another. For example, in the vapour phase sodium chloride exists as diatomic "molecules".

Brittleness

Most ionic compounds are very brittle. Once they reach the limit of their strength, they cannot deform mealleably, because the strict alignment of positive and negative ions must be maintained. Instead the material undergoes fracture via cleavage. As the temperature is elevated (usually close to the melting point) a ductile–brittle transition occurs, and plastic flow becomes possible by the motion of dislocations.

Compressibility

The compressibility of an ionic compound is strongly determined by its structure, and in particular the coordination number. For example, halides with the caesium chloride structure (coordination number 8) are less compressible than those with the sodium chloride structure (coordination number 6), and less again than those with a coordination number of 4.

Solubility

When ionic compounds dissolve, the individual ions dissociate and are solvated by the solvent and dispersed throughout the resulting solution. Because the ions are released into solution when dissolved, and can conduct charge, soluble ionic compounds are the most common class of strong electrolytes, and their solutions have a high electrical conductivity.

The solubility is highest in polar solvents (such as water) or ionic liquids, but tends to be low in nonpolar solvents (such as petrol/gasoline). This is principally because the resulting ion–dipole interactions are significantly stronger than ion-induced dipole interactions, so the heat of solution is higher. When the oppositely charged ions in the solid ionic lattice are surrounded by the opposite pole of a polar molecule, the solid ions are pulled out of the lattice and into the liquid. If the solvation energy exceeds the lattice energy, the negative net enthalpy change of solution provides a thermodynamic drive to remove ions from their positions in the crystal and dissolve in the liquid. In addition, the entropy change of solution is usually positive for most solid solutes like ionic compounds, which means that their solubility increases when the temperature increases. There are some unusual ionic compounds such as cerium(III) sulfate, where this entropy change is negative, due to extra order induced in the water upon solution, and the solubility decreases with temperature.

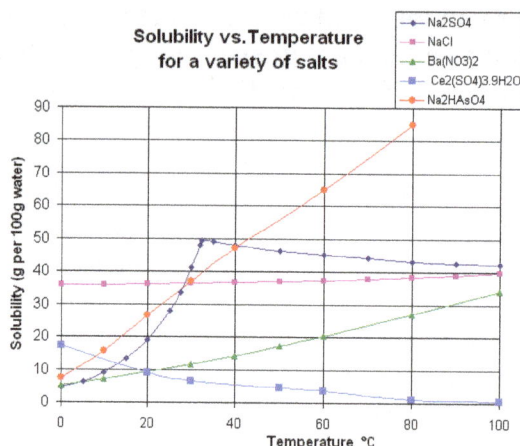

The aqueous solubility of a variety of ionic compounds as a function of temperature.
Some compounds exhibiting unusual solubility behaviour have been included.

Electrical Conductivity

Although ionic compounds contain charged atoms or clusters, these materials do not typically conduct electricity to any significant extent when the substance is solid. In order to conduct, the charged particles must be mobile rather than stationary in a crystal lattice. This is achieved to some degree at high temperatures when the defect concentration increases the ionic mobility and solid state ionic conductivity is observed. When the ionic compounds are dissolved in a liquid or are melted into a liquid, they can conduct electricity because the ions become completely mobile. This conductivity gain upon dissolving or melting is sometimes used as a defining characteristic of ionic compounds.

In some unusual ionic compounds: fast ion conductors, and ionic glasses, one or more of the ionic components has a significant mobility, allowing conductivity even while the material as a whole remains solid. This is often highly temperature dependant, and may be the result of either a phase change or a high defect concentration. These materials are used in all solid-state supercapacitors, batteries, and fuel cells, and in various kinds of chemical sensors.

Colour

Anhydrous cobalt(II) chloride,
$CoCl_2$

Cobalt(II) chloride hexahydrate,
$CoCl_2 \cdot 6H_2O$

The colour of an ionic compound is often different to the colour of an aqueous solution containing the constituent ions, or the hydrated form of the same compound.

The anions in compounds with bonds with the most ionic character tend to be colourless (with an absorption band in the ultraviolet part of the spectrum). In compounds with less ionic character, their colour deepens through yellow, orange, red and black (as the absorption band shifts to longer wavelengths into the visible spectrum).

The absorption band of simple cations shift toward shorter wavelength when they are involved in more covalent interactions. This occurs during hydration of metal ions, so colourless anhydrous ionic compounds with an anion absorbing in the infrared can become colourful in solution.

Uses

Ionic compounds have long had a wide variety of uses and applications. Many minerals are ionic. Humans have processed common salt (sodium chloride) for over 8000 years, using it first as a food seasoning and preservative, and now also in manufacturing, agriculture, water conditioning, for de-icing roads, and many other uses. Many ionic compounds are so widely used in society that they go by common names unrelated to their chemical identity. Examples of this include borax, calomel, milk of magnesia, muriatic acid, oil of vitriol, saltpeter, and slaked lime.

Soluble ionic compounds like salt can easily be dissolved to provide electrolyte solutions. This is a simple way to control the concentration and ionic strength. The concentration of solutes affects many colligative properties, including increasing the osmotic pressure, and causing freezing-point depression and boiling-point elevation. Because the solutes are charged ions they also increase the electrical conductivity of the solution. The increased ionic strength reduces the thickness of the electrical double layer around colloidal particles, and therefore the stability of emulsions and suspensions.

The chemical identity of the ions added is also important in many uses. For example, fluoride containing compounds are dissolved to supply fluoride ions for water fluoridation.

Solid ionic compounds have long been used as paint pigments, and are resistant to organic solvents, but are sensitive to acidity or basicity. Since 1801 pyrotechnicians have described and widely used metal-containing ionic compounds as sources of colour in fireworks. Under intense heat, the electrons in the metal ions or small molecules can be excited. These electrons later return to lower energy states, and release light with a colour spectrum characteristic of the species present.

In chemistry, ionic compounds are often used as precursors for high-temperature solid-state synthesis.

Many metals are geologically most abundant as ionic compounds within ores. To obtain the elemental materials, these ores are processed by smelting or electrolysis, in which redox reactions occur (often with a reducing agent such as carbon) such that the metal ions gain electrons to become neutral atoms.

Nomenclature

According to the nomenclature recommended by IUPAC, ionic compounds are named according to their composition, not their structure. In the most simple case of a binary ionic compound with no possible ambiguity about the charges and thus the stoichiometry, the common name is written using two words. The name of the cation (the unmodified element name for monatomic cations) comes first, followed by the name of the anion. For example, $MgCl_2$ is named magnesium chloride, and Na_2SO_4 is named sodium sulfate (SO_4^{2-} , sulfate, is an example of a polyatomic ion). To obtain the empirical formula from these names, the stoichiometry can be deduced from the charges on the ions, and the requirement of overall charge neutrality.

If there are multiple different cations and/or anions, multiplicative prefixes (*di-*, *tri-*, *tetra-*, ...) are often required to indicate the relative compositions, and cations then anions are listed in alphabetical order. For example, $KMgCl_3$ is named magnesium potassium trichloride to distinguish it from K_2MgCl_4, magnesium dipotassium tetrachloride (note that in both the empirical formula and the written name, the cations appear in alphabetical order, but the order varies between them because the symbol for potassium is K). When one of the ions already has a multiplicative prefix within its name, the alternate multiplicative prefixes (*bis-*, *tris-*, *tetrakis-*, ...) are used. For example, $Ba(BrF_4)_2$ is named barium bis(tetrafluoridobromate).

Compounds containing one or more elements which can exist in a variety of charge/oxidation states will have a stoichiometry that depends on which oxidation states are present, to ensure overall neutrality. This can be indicated in the name by specifying either the oxidation state of the elements present, or the charge on the ions. Because of the risk of ambiguity in allocating oxidation states, IUPAC prefers direct indication of the ionic charge numbers. These are written as an arabic integer followed by the sign (... , 2–, 1–, 1+, 2+, ...) in parentheses directly after the name of the cation (without a space separating them). For example, $FeSO_4$ is named iron(2+) sulfate (with the 2+ charge on the Fe^{2+} ions balancing the 2– charge on the sulfate ion), whereas $Fe_2(SO_4)_3$ is named iron(3+) sulfate (because the two iron ions in each formula unit each have a charge of 3+, to balance the 2– on each of the three sulfate ions). Stock nomenclature, still in common use, writes the oxidation number in Roman numerals (... , –II, –I, 0, I, II, ...). So the examples given above would be named iron(II) sulfate and iron(III) sulfate respectively. For simple ions the ionic charge and the oxidation number are identical, but for polyatomic ions they often differ. For example, the uranyl(2+) ion, UO_2^{2+} , has uranium in an oxidation state of +6, so would be called a dioxouranium(VI) ion in Stock nomenclature. An even older naming system for metal cations, also still widely used, appended the suffixes *-ous* and *-ic* to the Latin root of the name, to give special names for the low and high oxidation states. For example, this scheme uses "ferrous" and "ferric", for iron(II) and iron(III) respectively, so the examples given above were classically named ferrous sulfate and ferric sulfate.

Ionic Compounds to AB Type

(A) Zinc sulfide (ZnS) structure:

In ZnS, both cation (Zn) and anion (S) are in II oxidation state. The radius ratio in ZnS is 0.40 suggests a tetrahedral arrangement where each Zn^{2+} ion is surrounded by four S^{2-} anions. The co-ordination number of both ions is 4. Therefore, the arrangement is called 4:4 arrangement.

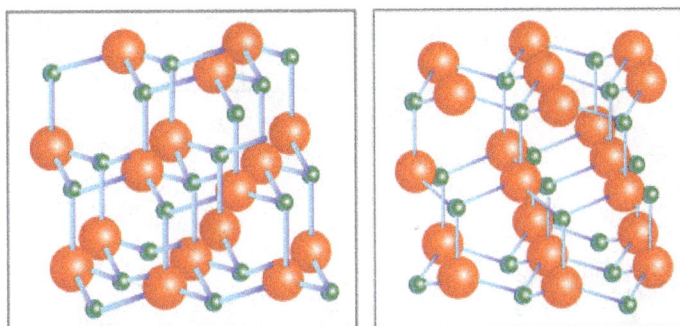

Zinc Blende   Wurtzite

Crystal structure of zinc blend and wurtzite.

(B) Sodium chloride (NaCl) structure:

In NaCl both cation (Na) and anion (Cl) are in I state. The radius ration 0.52 suggests an octahedral arrangement around the ions. The coordination is thus 6:6.

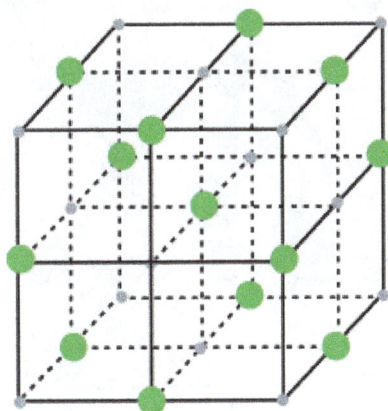

Cl^- Na^+

Crystal structure of NaCl.

(C) Caesium chlodire (CsCl) structure:

In CsCl both cation (Cs) and anion (Cl) are in I state. The radius ration 0.93 suggests an body centered cubic type arrangement around the ions. In this case each ion is surrounded by eight oppositely charges ions. The coordination is thus 8:8.

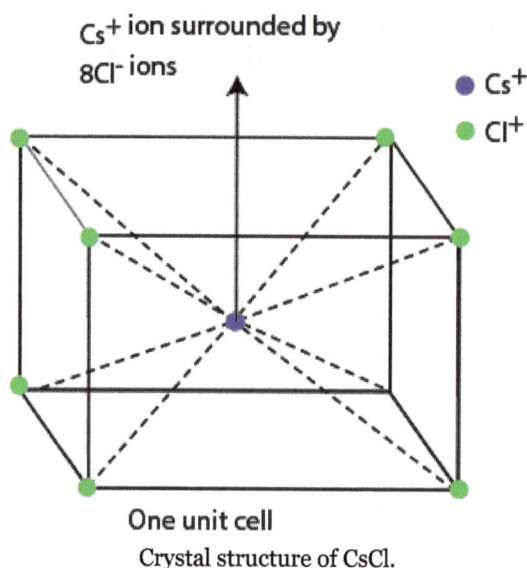

Cs^+ ion surrounded by 8Cl^- ions

● Cs^+
● Cl^+

One unit cell

Crystal structure of CsCl.

Ionic compounds to AB_2 type:

(A) Calcium fluoride (CaF_2, fluorite) structure:

In CaF_2 the radius ratio is 0.73 or above. It has a body centered cubic arrangement of fluoride ions around calcium ion. The ratio between Ca^{2+} and F^- is 2:1. Since, there is twice as many fluoride ions as calcium ion, the coordination number of the both ions will be different. A 8:4 F^-:Ca^{2+} arrangement has been found for CaF_2 where each Ca^{2+} ions are tetrahedrally arranged around each F^- ion.

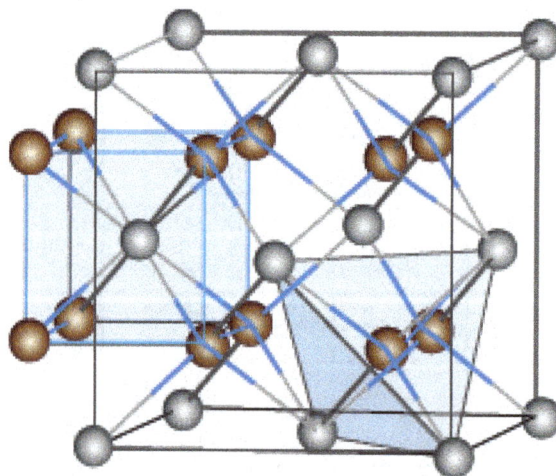

Crystal structure of CaF_2.

(B) TiO_2 (rutile) structure:

TiO_2 exists in three different forms; (i) anatase, (ii) brookite, and (iii) rutile. In many crystals rutile structure is found and the radius ration is between 0.41 and 0.73. This indicates a coordination number is 6. Therefore the other will be 3 (as it is AB_2 kind structure). Rutile form has a 3:6 structure where each Ti^{4+} is octahedrally surrounded by six O^{2-} ions and each O^{2-} ion has three Ti^{4+} ions around in a plane triangular fashion.

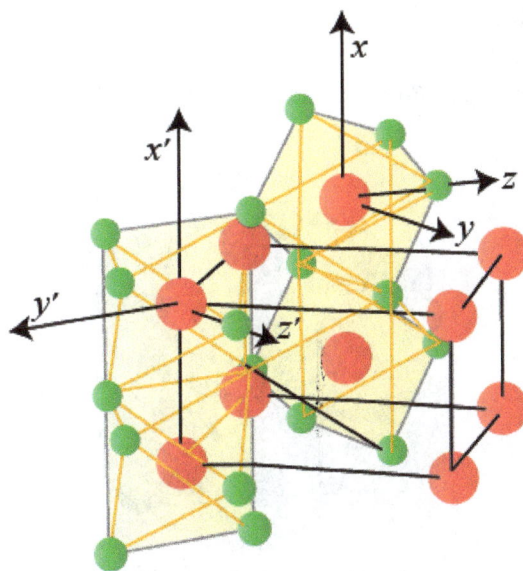

Crystal structure of TiO_2 (rutile).

The rutile structure is not close packed. The unit cell is not a cube as one of the three axes is about 30% shorter. It is convenient to describe as distorted cube.

Limitations of the radius ratio rule:

The structure adapted by the alkali metal halides may be considered to demonstrate the failure of structural prediction from the radius ratio rule.

LiCl, LiBr, and LiI, the radius ratio lies around 0.3 suggesting fourfold coordination. In case of KF, KCl, RbF, RbCl, RbBr, and CsF, the radius ratio is 0.73, hence coordination number shoud be minimum 8. In fact, all the alkali metal halides adapt 6:6 (NaCl type) coordination environment except CsCl, CsBr, and CsI.

It is said in the radius ratio rule that with increase of the ratio value the coordination number will increase. Let consider a AB type ionic compound, as we know that with increase of radius ration value the number of counter ions around a particular ion will increase. As the system is AB type,

therefore, each cation will be surrounded by one anion. So, the gain in electrostatic attraction force will be cancelled by large repulsion force by the same charge ions as the coordination number is high. Therefore, the stability of ionic solid cannot be answered.

This rule provides a rough guide to the structure of ionic solids. Ultimately, the reason why any particular crystal structure is formed is that it gives the most favorable lattice energy.

Lattice Energy

The lattice energy of a crystalline solid is usually defined as the energy of formation of the crystals from infinitely-separated ions and as such is invariably positive. The precise value of the lattice energy may not be determined experimentally, because of the impossibility of preparing an adequate amount of gaseous ions or atoms and measuring the energy released during their condensation to form the solid. However, the value of the lattice energy may either be derived theoretically from electrostatics or from a thermodynamic cycling reaction, the Born–Haber cycle.

Historical Development

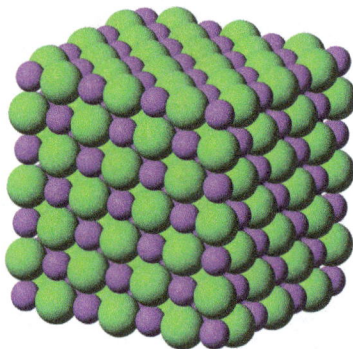

Sodium chloride crystal lattice

The concept of lattice energy was originally developed for rocksalt-structured and sphalerite-structured compounds like NaCl and ZnS, where the ions occupy high-symmetry crystal lattice sites. In the case of NaCl, lattice energy is the energy released by the reaction

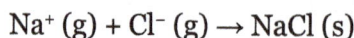

$$Na^+ (g) + Cl^- (g) \rightarrow NaCl (s)$$

which would amount to -786 kJ/mol.

Some older textbooks define lattice energy with the opposite sign, i.e. the energy required to convert the crystal into infinitely separated gaseous ions in vacuum, an endothermic process. Following this convention, the lattice energy of NaCl would be +786 kJ/mol. The lattice energy for ionic crystals such as sodium chloride, metals such as iron, or covalently linked materials such as diamond is considerably greater in magnitude than for solids such as sugar or iodine, whose neutral molecules interact only by weaker dipole-dipole or van der Waals forces.

Theoretical Treatments

The relationship between the molar lattice energy and the molar lattice enthalpy is given by the following equation:

$$\Delta_G U = \Delta_G H - p \Delta V_m,$$

where $\Delta_G U$ is the molar lattice energy, $\Delta_G H$ the molar lattice enthalpy and ΔV_m the change of the volume per mol. Therefore, the lattice enthalpy further takes into account that work has to be performed against an outer pressure p. Lattice Energy of an ionic compound depends upon charge of the ion and size of the ions. Moreover, factors such as packing of ions doesn't matter efficiently.

Born–Landé Equation

In 1918 Born and Landé proposed that the lattice energy could be derived from the electric potential of the ionic lattice and a repulsive potential energy term.

$$E = -\frac{N_A M z^+ z^- q_e^2}{4\pi\varepsilon_0 r_0}\left(1-\frac{1}{n}\right),$$

where

N_A is the Avogadro constant;

M is the Madelung constant, relating to the geometry of the crystal;

z^+ is the charge number of cation;

z^- is the charge number of anion;

q_e is the elementary charge, equal to 1.6022×10^{-19} C;

ε_0 is the permittivity of free space, equal to 8.854×10^{-12} C^2 J^{-1} m^{-1};

r_0 is the distance to closest ion; and

n is the Born exponent, a number between 5 and 12, determined experimentally by measuring the compressibility of the solid, or derived theoretically.

The Born–Landé equation gives a reasonable fit to the lattice energy.

Compound	Calculated Lattice Energy	Experimental Lattice Energy
NaCl	−756 kJ/mol	−787 kJ/mol
LiF	−1007 kJ/mol	−1046 kJ/mol
$CaCl_2$	−2170 kJ/mol	−2255 kJ/mol

From the Born–Landé equation it can be seen that the lattice energy of a compound is dependent on a number of factors

- as the charges on the ions increase the lattice energy increases (becomes more negative),

- when ions are closer together the lattice energy increases (becomes more negative)

Barium oxide (BaO), for instance, which has the NaCl structure and therefore the same Madelung constant, has a bond radius of 275 picometers and a lattice energy of -3054 kJ/mol, while sodium chloride (NaCl) has a bond radius of 283 picometers and a lattice energy of -786 kJ/mol.

Kapustinskii Equation

The Kapustinskii equation can be used as a simpler way of deriving lattice energies where high precision is not required.

Effect of Polarisation

For ionic compounds with ions occupying lattice sites with crystallographic point groups C_1, C_{1h}, C_n or C_{nv} ($n = 2, 3, 4$ or 6) the concept of the lattice energy and the Born–Haber cycle has to be extended. In these cases the polarization energy E_{pol} associated with ions on polar lattice sites has to be included in the Born–Haber cycle and the solid formation reaction has to start from the already polarized species. As an example, one may consider the case of iron-pyrite FeS_2, where sulfur ions occupy lattice site of point symmetry group C_3. The lattice energy defining reaction then reads

$$Fe^{2+} (g) + 2 \text{ pol } S^- (g) \rightarrow FeS_2 (s)$$

where pol S^- stands for the polarized, gaseous sulfur ion. It has been shown that the neglection of the effect led to 15% difference between theoretical and experimental thermodynamic cycle energy of FeS_2 that reduced to only 2%, when the sulfur polarization effects were included.

Lattice Defects

Why does not $NaCl_2$ exist?

The composition of $NaCl_2$ will be Na^{2+} and $2Cl^-$.

Process		Related energy terms
$Na(s)$	$Na(g)$	$+S$ (enthalpy of atomization)
$Na(g)$	$Na^{2+}(g)$	$+I = +(I_1+I_2)$ (I_1 and I_2 are respective first and second ionization energy)
$Cl_2(g)$	$2Cl(g)$	D (enthalpy of bond dissociation)
$2Cl(g)$	$2Cl^-(g)$	$-2E.A$ (electron affinity)
$Na^{2+}(g) + 2Cl^-(g)$	$NaCl_2(s)$	$-U_0$ (lattice energy)
$Na(s) + Cl_2(g)$	$NaCl_2(s)$	$+S+I+ D-2E.A-U_0 = -H_f$ (heat of formation of NaCl) H_f is negative as heat release on the formation of NaCl.

Therefore, we can write;

$$+S+(I_1+I_2)+D-2E.A-U_0=H_f$$

$NaCl_2$ would crystallize in fluorite (CaF_2) structure with $A = 2.52$. Theoretically, the U_0 for $NaCl_2$ obtained is 2180 kJmol^{-1}.

Hence,

$$H_f = +S+(I_1+I_2)+D-2E.A-U_0$$
$$= +108+(496+5462)+242-2\times348-2180$$
$$= 2530\,kJmol^{-1}$$

As the H_f is positive, hence $NaCl_2$ cannot be stabilized. Here, the very large I_2 (5462 kJmol^{-1}) of Na cannot be compensate by the stabilizing lattice energy.

Why does not CaCl exist?

Process		Related energy terms
Ca(s)	Ca(g)	+S (enthalpy of atomization)
Ca(g)	Ca⁺(g)	+I (ionization energy)
½ F(g)	F(g)	½ D (enthalpy of bond dissociation)
F(g)	F⁻(g)	E.A (electron affinity)
Ca²⁺(g) + Cl⁻(g)	CaF(s)	−U₀ (lattice energy)
Na(s) + ½ F₂(g)	NaF(s)	$+S + I + \frac{1}{2} D - E.A - U_0 = -H_f$ (heat of formation of NaCl)
		H_f is negative as heat release on the formation of NaCl.

Assume that CaF will be crystallized in the same geometry as KF. Therefore, the calculated lattice energy will be $U_0 = -795$ kJmol⁻¹.

$$\text{Hence, } H_f = +S + I + 1/2\,D - E.A - U_0$$
$$= +178 + 590 + 79 - 328 - 795$$
$$= -276\,\text{kJmol}^{-1}$$

As the H_f value is not large (276 kJmol⁻¹), the CaF is unstable and disproportionate to CaF_2 and Ca.

Lattice defects:

In reality, lattices do not acquire a perfectly ordered arrangement of the constituting ions. A defective lattice might not conform to the stoichiometric cation and anion ratio as represented by its chemical formula. Hence, lattices can be divided in (i) stoichiometric defect and (ii) non- stoichiometric defect.

Stoichiometric defect:

Frenkel defect and Schottky defect belong to this type of lattice defect.

Frenkel defect:

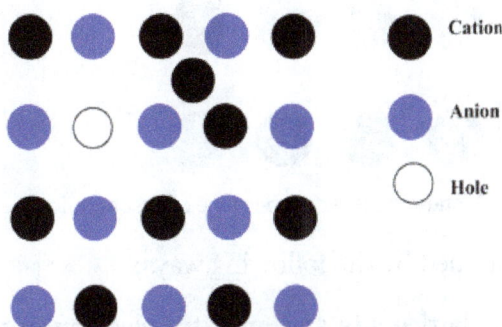

In Frenkel defect, an ion is missing from its lattice site. Therefore, a hole is generated in the lattice. The missing ion occupies some interstitial position. Crystals with small positive ions and large negative ion, like AgCl, AgBr, AgI, ZnS, etc shows this kind of defect.

Schottky defect:

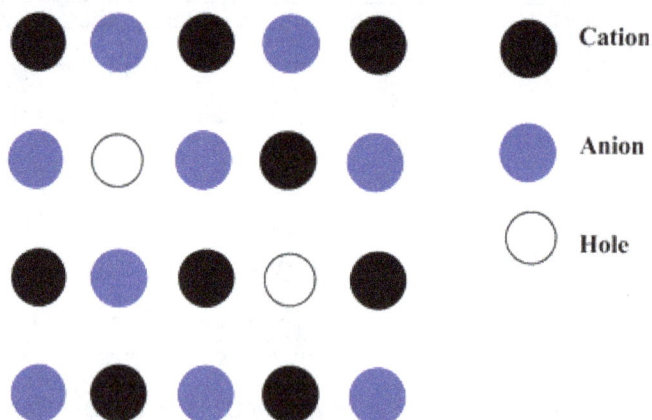

In Schottky defect, both a positive ion and a negative ion are missing from its lattice site. Therefore, two holes are generated in the lattice. Some examples are NaCl, KCl, KBr etc.

Crystalline solid with these kinds of defect show some sort of electric conductivity via ion migration to the hole(s).

Non-stoichiometric Defects:

In the non-stoichiometric defect the ratio of cation and anion differ from that of the indicated chemical formula. The charge is balanced is either by extra electron or by cation according to necessary. In this type of compounds, either the matal or the non-metal atoms may be in excess.

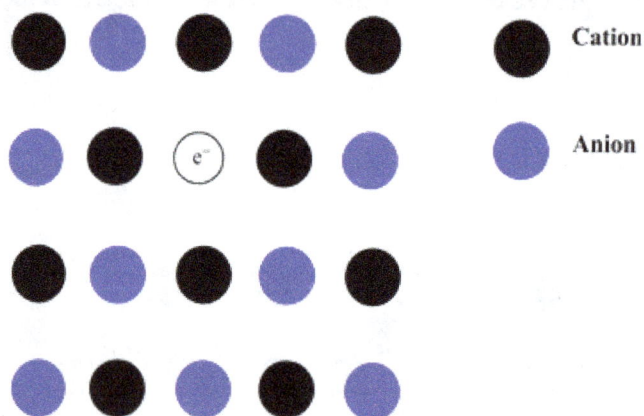

Metal-excess defect because of absent anion.

The excess of metal can be obtained by the following ways;

(A) Missing of anion from the lattice site and an extra electron in present there to balance the charge.

Example: NaCl treated with Na-vapor forms such a yellow non-stoichiometric lattice. KCl with K-vapour becomes bule.

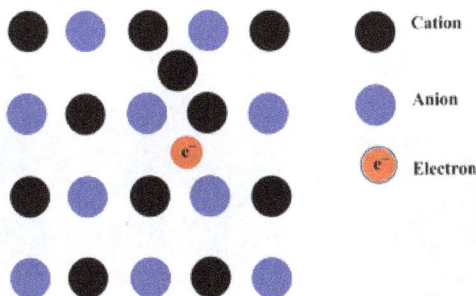

Metal excess defect because of excess of cation(s).

(B) Presence of an extra metal ion in interstitial position and extra electron in some other interstitial position to balance the charge of the lattice.

Example: ZnO, CdO, Fe_2O_3, etc shows such type of lattice defect. ZnO under normal room temperature is white in color but when it is heated it becomes yellow. This is because it the presence of extra electrons, at the interstitial position of ZnO lattice, undergo to higher energy excited states by absorbing heat.

The metal deficiency can be obtained by the following ways;

(C) Positive ion may absent in a lattice and a doubly charged cation maintains the charge balance.

Example: FeO, CuO, NiO, Fes, etc.

Metal deficiency because of mission cation.

(D) A non-stoichiometric defect might be possible as shown in case (C) with a additional anion in the interstitial position. No example of crystals containing this kind of anion in the interstitial position of a lattice is known.

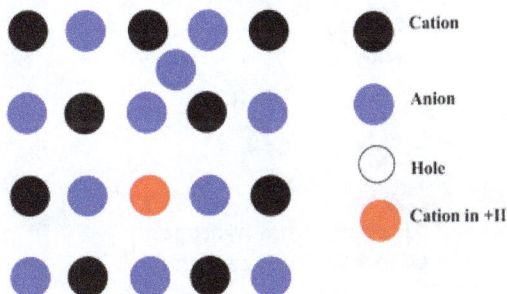

Metal deficiency because of the presence of interstitial anion.

Covalent Bond

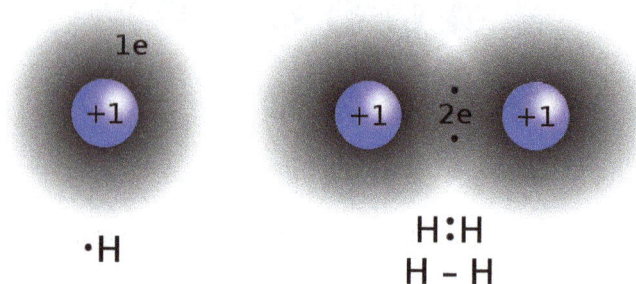

A covalent bond forming H_2 (right) where two hydrogen atoms share the two electrons

A covalent bond, also called a molecular bond, is a chemical bond that involves the sharing of electron pairs between atoms. These electron pairs are known as shared pairs or bonding pairs, and the stable balance of attractive and repulsive forces between atoms, when they share electrons, is known as covalent bonding. For many molecules, the sharing of electrons allows each atom to attain the equivalent of a full outer shell, corresponding to a stable electronic configuration.

Covalent bonding includes many kinds of interactions, including σ-bonding, π-bonding, metal-to-metal bonding, agostic interactions, bent bonds, and three-center two-electron bonds. The term *covalent bond* dates from 1939. The prefix *co-* means *jointly, associated in action, partnered to a lesser degree,* etc.; thus a "co-valent bond", in essence, means that the atoms share "valence", such as is discussed in valence bond theory.

In the molecule H_2, the hydrogen atoms share the two electrons via covalent bonding. Covalency is greatest between atoms of similar electronegativities. Thus, covalent bonding does not necessarily require that the two atoms be of the same elements, only that they be of comparable electronegativity. Covalent bonding that entails sharing of electrons over more than two atoms is said to be delocalized.

History

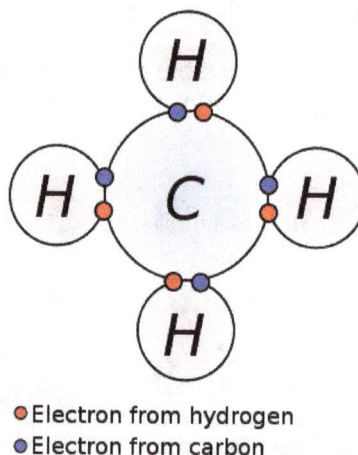

● Electron from hydrogen
● Electron from carbon

Early concepts in covalent bonding arose from this kind of image of the molecule of methane. Covalent bonding is implied in the Lewis structure by indicating electrons shared between atoms.

The term *covalence* in regard to bonding was first used in 1919 by Irving Langmuir in a *Journal of the American Chemical Society* article entitled "The Arrangement of Electrons in Atoms and Molecules". Langmuir wrote that "we shall denote by the term *covalence* the number of pairs of electrons that a given atom shares with its neighbors."

The idea of covalent bonding can be traced several years before 1919 to Gilbert N. Lewis, who in 1916 described the sharing of electron pairs between atoms. He introduced the *Lewis notation* or *electron dot notation* or *Lewis dot structure*, in which valence electrons (those in the outer shell) are represented as dots around the atomic symbols. Pairs of electrons located between atoms represent covalent bonds. Multiple pairs represent multiple bonds, such as double bonds and triple bonds. An alternative form of representation, not shown here, has bond-forming electron pairs represented as solid lines.

Lewis proposed that an atom forms enough covalent bonds to form a full (or closed) outer electron shell. In the diagram of methane shown here, the carbon atom has a valence of four and is, therefore, surrounded by eight electrons (the octet rule), four from the carbon itself and four from the hydrogens bonded to it. Each hydrogen has a valence of one and is surrounded by two electrons (a duet rule) – its own one electron plus one from the carbon. The numbers of electrons correspond to full shells in the quantum theory of the atom; the outer shell of a carbon atom is the $n = 2$ shell, which can hold eight electrons, whereas the outer (and only) shell of a hydrogen atom is the $n = 1$ shell, which can hold only two.

While the idea of shared electron pairs provides an effective qualitative picture of covalent bonding, quantum mechanics is needed to understand the nature of these bonds and predict the structures and properties of simple molecules. Walter Heitler and Fritz London are credited with the first successful quantum mechanical explanation of a chemical bond (molecular hydrogen) in 1927. Their work was based on the valence bond model, which assumes that a chemical bond is formed when there is good overlap between the atomic orbitals of participating atoms.

Types of Covalent Bonds

Atomic orbitals (except for s orbitals) have specific directional properties leading to different types of covalent bonds. Sigma (σ) bonds are the strongest covalent bonds and are due to head-on overlapping of orbitals on two different atoms. A single bond is usually a σ bond. Pi (π) bonds are weaker and are due to lateral overlap between p (or d) orbitals. A double bond between two given atoms consists of one σ and one π bond, and a triple bond is one σ and two π bonds.

Covalent bonds are also affected by the electronegativity of the connected atoms which determines the chemical polarity of the bond. Two atoms with equal electronegativity will make nonpolar covalent bonds such as H–H. An unequal relationship creates a polar covalent bond such as with H–Cl. However polarity also requires geometric asymmetry, or else dipoles may cancel out resulting in a non-polar molecule.

Covalent Structures

There are several types of structures for covalent substances, including individual molecules, molecular structures, macromolecular structures and giant covalent structures. Individual molecules have strong bonds that hold the atoms together, but there are negligible forces of attraction between molecules. Such covalent substances are usually gases, for example, HCl, SO_2, CO_2, and CH_4. In molecular structures, there are weak forces of attraction. Such covalent substances are low-boiling-temperature liquids (such as ethanol), and low-melting-temperature solids (such as iodine and solid CO_2). Macromolecular structures have large numbers of atoms linked by covalent bonds in chains, including synthetic polymers such as polyethylene and nylon, and biopolymers such as proteins and starch. Network covalent structures (or giant covalent structures) contain large numbers of atoms linked in sheets (such as graphite), or 3-dimensional structures (such as diamond and quartz). These substances have high melting and boiling points, are frequently brittle, and tend to have high electrical resistivity. Elements that have high electronegativity, and the ability to form three or four electron pair bonds, often form such large macromolecular structures.

One- and Three-electron Bonds

Bonds with one or three electrons can be found in radical species, which have an odd number of electrons. The simplest example of a 1-electron bond is found in the dihydrogen cation, H_2^+. One-electron bonds often have about half the bond energy of a 2-electron bond, and are therefore called "half bonds". However, there are exceptions: in the case of dilithium, the bond is actually stronger for the 1-electron Li_2^+ than for the 2-electron Li_2. This exception can be explained in terms of hybridization and inner-shell effects.

2e bond (e.g., CH_4)

3e bond (e.g., NO)

Comparison of the electronic structure of the three-electron bond to the conventional covalent bond.

The simplest example of three-electron bonding can be found in the helium dimer cation, He_2^+. It is considered a "half bond" because it consists of only one shared electron (rather than two); in molecular orbital terms, the third electron is in an anti-bonding orbital which cancels out half of the bond formed by the other two electrons. Another example of a molecule containing a 3-electron bond, in addition to two 2-electron bonds, is nitric oxide, NO. The oxygen molecule, O_2 can also be regarded as having two 3-electron bonds and one 2-electron bond, which accounts for its paramagnetism and its formal bond order of 2. Chlorine dioxide and its heavier analogues bromine dioxide and iodine dioxide also contain three-electron bonds.

Molecules with odd-electron bonds are usually highly reactive. These types of bond are only stable between atoms with similar electronegativities.

Resonance

There are situations whereby a single Lewis structure is insufficient to explain the electron configuration in a molecule, hence a superposition of structures are needed. The same two atoms in such molecules can be bonded differently in different structures (a single bond in one, a double bond in another, or even none at all), resulting in a non-integer bond order. The nitrate ion is one such example with three equivalent structures. The bond between the nitrogen and each oxygen is a double bond in one structure and a single bond in the other two, so that the average bond order for each N–O interaction is $\dfrac{2+1+1}{3}=\dfrac{4}{3}$.

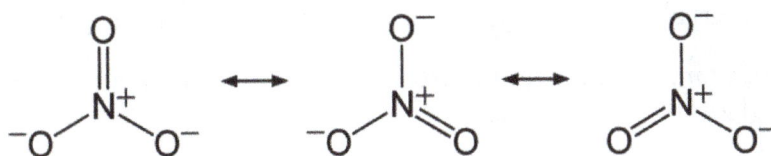

Aromaticity

In organic chemistry, when a molecule with a planar ring obeys Hückel's rule, where the number of π electrons fit the formula $4n + 2$ (where n is an integer), it attains extra stability and symmetry. In benzene, the prototypical aromatic compound, there are 6 π bonding electrons ($n = 1$, $4n + 2 = 6$). These occupy three delocalized π molecular orbitals (molecular orbital theory) or form conjugate π bonds in two resonance structures that linearly combine (valence bond theory), creating a regular hexagon exhibiting a greater stabilization than the hypothetical 1,3,5-cyclohexatriene.

In the case of heterocyclic aromatics and substituted benzenes, the electronegativity differences between different parts of the ring may dominate the chemical behaviour of aromatic ring bonds, which otherwise are equivalent.

Hypervalence

Certain molecules such as xenon difluoride and sulfur hexafluoride have higher co-ordination numbers than would be possible due to strictly covalent bonding according to the octet rule. This is explained by the three-center four-electron bond ("3c–4e") model which interprets the molecular wavefunction in terms of non-bonding highest occupied molecular orbitals in molecular orbital theory and ionic-covalent resonance in valence bond theory.

Electron-deficiency

In three-center two-electron bonds ("3c–2e") three atoms share two electrons in bonding. This type of bonding occurs in electron deficient compounds like diborane. Each such bond (2 per molecule in diborane) contains a pair of electrons which connect the boron atoms to each other in a banana shape, with a proton (nucleus of a hydrogen atom) in the middle of the bond, sharing electrons with both boron atoms. In certain cluster compounds, so-called four-center two-electron bonds also have been postulated.

Quantum Mechanical Description

After the development of quantum mechanics, two basic theories were proposed to provide a quantum description of chemical bonding: valence bond (VB) theory and molecular orbital (MO) theory. A more recent quantum description is given in terms of atomic contributions to the electronic density of states.

Covalency from Atomic Contribution to the Electronic Density of States

In COOP, COHP and BCOOP, evaluation of bond covalency is dependent on the basis set. To overcome this issue, an alternative formulation of the bond covalency can be provided in this way.

The center mass $cm(n,l,m_l,m_s)$ of an atomic orbital $|n,l,m_l,m_s\rangle$, with quantum numbers n, l, m_l, m_s, for atom A is defined as

$$cm^A(n,l,m_l,m_s) = \frac{\int_{E_0}^{E_1} E g^A_{|n,l,m_l,m_s\rangle}(E)\,dE}{\int_{E_0}^{E_1} g^A_{|n,l,m_l,m_s\rangle}(E)\,dE}$$

where $g^A_{|n,l,m_l,m_s\rangle}(E)$ is the contribution of the atomic orbital $|n,l,m_l,m_s\rangle$ of the atom A to the total electronic density of states $g(E)$ of the solid

$$g(E) = \sum_{A} \sum_{n,l} \sum_{m_l,m_s} g^A_{|n,l,m_l,m_s\rangle}(E)$$

where the outer sum runs over all atoms A of the unit cell. The energy window $[E_0, E_1]$ is chosen in such a way that it encompasses all relevant bands participating in the bond. If the range to select is unclear, it can be identified in practice by examining the molecular orbitals that describe the electron density along the considered bond.

The relative position $C_{n_A l_A, n_B l_B}$ of the center mass of $|n_A, l_A\rangle$ levels of atom A with respect to the center mass of $|n_B, l_B\rangle$ levels of atom B is given as

$$C_{n_A l_A, n_B l_B} = -\left| cm^A(n_A, l_A) - cm^B(n_B, l_B) \right|$$

where the contributions of the magnetic and spin quantum numbers are summed. According to this definition, the relative position of the A levels with respect to the B levels is

$$C_{A,B} = -\left| cm^A - cm^B \right|$$

where, for simplicity, we may omit the dependence from the principal quantum number n in the notation referring to $C_{n_A l_A, n_B l_B}$.

In this formalism, the greater the value of $C_{A,B}$, the higher the overlap of the selected atomic bands, and thus the electron density described by those orbitals gives a more covalent A–B bond. The quantity $C_{A,B}$ is denoted as the *covalency* of the A–B bond, which is specified in the same units of the energy E.

A covalent bond is a chemical bonding formed by sharing of pairs of electrons between two atoms.

Lewis Theory (Octet rule):

Rule: A stable arrangement is attended when the atom is surrounded by eight electrons. This octet can be made up by own electrons and some electrons which are shared. Thus, an atom continues to form bonds until an octet of electrons is made. This is known as octet rule by Lewis.

Example:

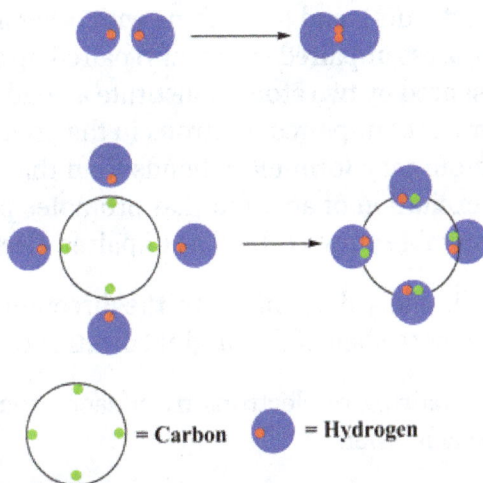

= Carbon   = Hydrogen

(i) Normally two electrons pairs up and forms a bond. Ex: H_2

(ii) For most atoms there will be a maximum of eight electrons in the valence shell (octet structure). Ex: CH_4

Exception of the octet rule:

(A) PF 3 maintains octet rule, while PF 5 does not. PF 5 has ten outer electrons.

(B) Molecules comprised of odd number of electrons do not follow octet rule. Ex: NO, ClO_2 , etc. This rule even does not explain the origin of paramagnetic character in O_2 molecule.

(C) For atoms like Be and B which have less than four outer electrons. Even if all the outer electrons used to form bonds an octet will not be resulted for Be and B.

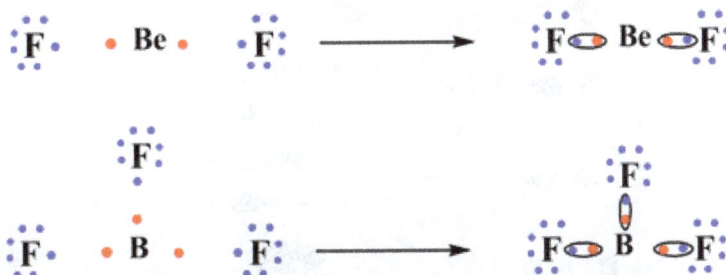

The octet rule has been further extended to include the exceptional part. Hence, the following rules are added to the octet rules.

(iii) For elements with available d orbitals, the valence shell can be expanded beyond an octet.

This rule explained the formation of PF 5. Here P has low-energy d orbitals.

(iv) The molecule will seek the lowest overall energy. This means the maximum number of bonds will form, that the strongest possible bonds will form, and that the arrangement of the atoms in molecule will be in such as to minimize adverse repulsion energy.

Valence Bond Theory:

Atoms containing unpaired electron(s) tend to combine with other atom(s) which also possess(es) unpaired electron(s). In this process unpaired electron(s) paired up and attained a stable noble gas configuration. Two electrons shared by two atoms constitute a bond. The number of bonds formed by an atom is same as the number of unpaired electrons in the ground state (lowest energy state). However, in some cases the atom may form more bonds than that of its acquired unpaired electron. This takes place by the excitation of an atom that promotes paired-electrons in the ground state to the excited state in a suitable empty orbital as unpaired electron.

The shape of a molecule is preliminary determined by the direction in which the participating orbitals point. Electrons in the valence shell of the original orbital are known as lone pairs.

A covalent bond is resulted by pairing of electrons from each atom. In this case the spin of the electrons must be opposite to each other.

Limitations:

(i) Relative stability of different molecules cannot be explained.

(ii) Paramagnetic nature of complexes or molecules cannot be explained.

(iii) Different shape of molecules cannot be explained.

Hybridization:

Hybridization is the mixing of atomic orbitals prior to overlap that change the partial distribution of orbitals.

Due to this hybridization mainly two different kind of bonds forms, (i) sigma (σ) and (ii) pi (π) bond.

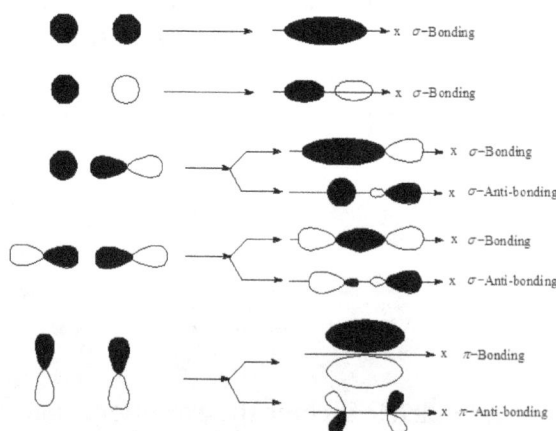

Sigma (σ) and pi (π) bonding and anti - bonding by the combination of s and p orbitals.

Sigma (σ) bond: A covalent bond established between two atoms having the maximum density of the electron cloud along the axis connecting the centers of the two participating atoms is called sigma (σ) bond.

Pi (π) bond: A bond is formed by the lateral overlap between two atomic orbitals possessing maximum electron density on both sides of the overlapping axis is known as pi (π) bond.

Therefore, from the definition it is clear that sigma (σ) bonds are more strong than pi (π) bonds.

Sigma (σ) bonds may arises from the overlap between s, p, and d orbitals, like (i) s - s orbitals, (ii) s - p prbitals, (iii) p - p orbitals, (iv) s - d orbitals, so on.

s orbitals are spherical, while p , d orbitals are dumbbell shaped. Hence, s - s over lap is weaker compared to s - p , p - p overlaps.

**Herein, it is important to note that due to hybridization of two atomic orbitals two molecular orbitals formed. One is called bonding orbital and the other is called anti - bonding orbital. Hence, n number of atomic orbitals forms n /2 bonding and n /2 anti - bonding orbitals.

Examples:

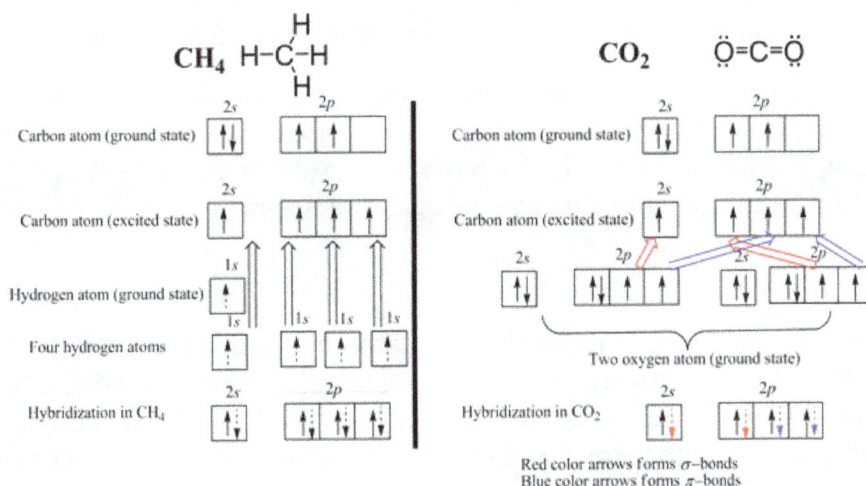

Red color arrows forms σ–bonds
Blue color arrows forms π–bonds

In the formation of CH_4 one s orbital and three p orbital of the central carbon atom participate.

Similarly, in the CO_2 molecule one s orbital and three p orbital of the central carbon atom participate but here two p orbitals of carbon atom forms two π - bonds. Therefore, CH_4 is sp^3 hybridized and CO_2 is sp^2 hybridized. NOTE: Orbitals participate in π - bond formation is not counted to the hybridization.

VSEPR Theory

Valence shell electron pair repulsion (VSEPR) theory is a model used in chemistry to predict the geometry of individual molecules from the number of electron pairs surrounding their central atoms. It is also named the Gillespie-Nyholm theory after its two main developers. The acronym "VSEPR" is pronounced either "*ves*-pur" or "vuh-*seh*-per."

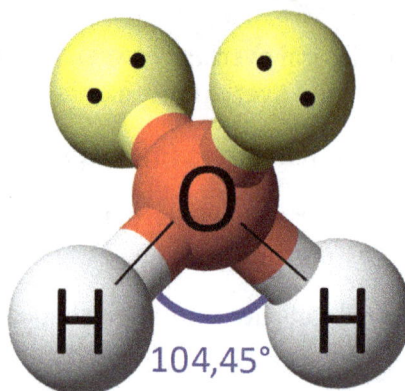

Example of bent electron arrangement. Shows location of unpaired electrons, bonded atoms, and bond angles. (Water molecule) The bond angle for water is 104.5°.

The premise of VSEPR is that the valence electron pairs surrounding an atom tend to repel each other and will, therefore, adopt an arrangement that minimizes this repulsion, thus determining the molecule's geometry. Gillespie has emphasized that the electron-electron repulsion due to the Pauli exclusion principle is more important in determining molecular geometry than the electrostatic repulsion.

VSEPR theory is based on observable electron density rather than mathematical wave functions and hence unrelated to orbital hybridisation, although both address molecular shape. While it is mainly qualitative, VSEPR has a quantitative basis in quantum chemical topology (QCT) methods such as the electron localization function and the quantum theory of atoms in molecules (QTAIM).

History

The idea of a correlation between molecular geometry and number of valence electrons (both shared and unshared) was originally proposed in 1939 by Ryutaro Tsuchida in Japan, and was independently presented in a Bakerian Lecture in 1940 by Nevil Sidgwick and Herbert Powell of the University of Oxford. In 1957, Ronald Gillespie and Ronald Sydney Nyholm of University College London refined this concept into a more detailed theory, capable of choosing between various alternative geometries.

Overview

VSEPR theory is used to predict the arrangement of electron pairs around non-hydrogen atoms in molecules, especially simple and symmetric molecules, where these key, central atoms participate in bonding to two or more other atoms; the geometry of these key atoms and their non-bonding electron pairs in turn determine the geometry of the larger whole.

The number of electron pairs in the valence shell of a central atom is determined after drawing the Lewis structure of the molecule, and expanding it to show all bonding groups and lone pairs of electrons. In VSEPR theory, a double bond or triple bond are treated as a single bonding group. The sum of the number of atoms bonded to a central atom and the number of lone pairs formed by its nonbonding valence electrons is known as the central atom's steric number.

The electron pairs (or groups if multiple bonds are present) are assumed to lie on the surface of a sphere centered on the central atom and tend to occupy positions that minimize their mutual repulsions by maximizing the distance between them. The number of electron pairs (or groups), therefore, determines the overall geometry that they will adopt. For example, when there are two electron pairs surrounding the central atom, their mutual repulsion is minimal when they lie at opposite poles of the sphere. Therefore, the central atom is predicted to adopt a *linear* geometry. If there are 3 electron pairs surrounding the central atom, their repulsion is minimized by placing them at the vertices of an equilateral triangle centered on the atom. Therefore, the predicted geometry is *trigonal*. Likewise, for 4 electron pairs, the optimal arrangement is *tetrahedral*.

Degree of Repulsion

The overall geometry is further refined by distinguishing between *bonding* and *nonbonding* electron pairs. The bonding electron pair shared in a sigma bond with an adjacent atom lies further from the central atom than a nonbonding (lone) pair of that atom, which is held close to its positively charged nucleus. VSEPR theory therefore views repulsion by the lone pair to be greater than the repulsion by a bonding pair. As such, when a molecule has 2 interactions with different degrees of repulsion, VSEPR theory predicts the structure where lone pairs occupy positions that allow them to experience less repulsion. Lone pair–lone pair (lp–lp) repulsions are considered stronger than lone pair–bonding pair (lp–bp) repulsions, which in turn are considered stronger than bonding pair–bonding pair (bp–bp) repulsions, distinctions that then guide decisions about overall geometry when 2 or more non-equivalent positions are possible. For instance, when 5 valence electron pairs surround a central atom, they adopt a trigonal bipyramidal molecular geometry with two collinear *axial* positions and three *equatorial* positions. An electron pair in an axial position has three close equatorial neighbors only 90° away and a fourth much farther at 180°, while an equatorial electron pair has only two adjacent pairs at 90° and two at 120°. The repulsion from the close neighbors at 90° is more important, so that the axial positions experience more repulsion than the equatorial positions; hence, when there are lone pairs, they tend to occupy equatorial positions.

The difference between lone pairs and bonding pairs may also be used to rationalize deviations from idealized geometries. For example, the H_2O molecule has four electron pairs in its valence shell: two lone pairs and two bond pairs. The four electron pairs are spread so as to point roughly towards the apices of a tetrahedron. However, the bond angle between the two O–H bonds is only 104.5°, rather than the 109.5° of a regular tetrahedron, because the two lone pairs (whose density or probability envelopes lie closer to the oxygen nucleus) exert a greater mutual repulsion than the two bond pairs.

An advanced-level explanation replaces the above distinction with two rules:

- Bent's rule: An electron pair of a more electropositive ligand constitutes greater repulsion. This explains why the Cl in $PClF_4$ prefers the equatorial position and why the bond angle in oxygen difluoride (103.8°) is smaller than that of water (104.5°). Lone pairs are then considered to be a special case of this rule, held by a "ghost ligand" in the limit of electropositivity.

- A higher bond order constitutes greater repulsion. This explains why in phosgene, the oxygen–chlorine bond angle (124.1°) is larger than the chlorine–chlorine bond angle (111.8°) even though chlorine is more electropositive than oxygen. In the carbonate ion, all three bond angles are equivalent due to resonance.

AXE Method

The "AXE method" of electron counting is commonly used when applying the VSEPR theory. The A represents the central atom and always has an implied subscript one. The X represents each of ligands (atoms bonded to A). The E represents the number of lone electron *pairs* surrounding the central atom. The sum of X and E is known as the steric number.

Based on the steric number and distribution of Xs and Es, VSEPR theory makes the predictions in the following tables. Note that the geometries are named according to the atomic positions only and not the electron arrangement. For example, the description of AX_2E_1 as a bent molecule means that the three atoms AX_2 are not in one straight line, although the lone pair helps to determine the geometry.

Steric number	Molecular geometry 0 lone pairs	Molecular geometry 1 lone pair	Molecular geometry 2 lone pairs	Molecular geometry 3 lone pairs
2	X—A—X Linear (CO_2)			
3	Trigonal planar (BCl_3)	V-shape (SO_2)		
4	Tetrahedral (CH_4)	Trigonal pyramidal (NH_3)	Bent (H_2O)	
5	Trigonal bipyramidal (PCl_5)	Seesaw (SF_4)	T-shaped (ClF_3)	Linear (I– 3)

| 6 | Octahedral (SF$_6$) | Square pyramidal (BrF$_5$) | Square planar (XeF$_4$) | |
| 7 | Pentagonal bipyramidal (IF$_7$) | Pentagonal pyramidal (XeOF$_5^-$) | Pentagonal planar (XeF$_5^-$) | |

Molecule type	Shape	Electron arrangement including lone pairs, shown in pale yellow	Geometry excluding lone pairs	Examples
AX$_2$E$_0$	Linear			BeCl$_2$, HgCl$_2$, CO$_2$
AX$_2$E$_1$	Bent			NO_2^- , SO$_2$, O$_3$, CCl$_2$
AX$_2$E$_2$	Bent			H$_2$O, OF$_2$
AX$_2$E$_3$	Linear			XeF$_2$, I$_3^-$, XeCl$_2$
AX$_3$E$_0$	Trigonal planar			BF$_3$, CO_3^{2-} , NO_3^- , SO$_3$

AX_6E_0	Octahedral				SF_6, WCl_6
AX_6E_1	Pentagonal pyramidal				$XeOF_5^-$, IOF_5^{2-}

When the substituent (X) atoms are not all the same, the geometry is still approximately valid, but the bond angles may be slightly different from the ones where all the outside atoms are the same. For example, the double-bond carbons in alkenes like C_2H_4 are AX_3E_0, but the bond angles are not all exactly 120°. Likewise, $SOCl_2$ is AX_3E_1, but because the X substituents are not identical, the X–A–X angles are not all equal.

As a tool in predicting the geometry adopted with a given number of electron pairs, an often used physical demonstration of the principle of minimal electron pair repulsion utilizes inflated balloons. Through handling, balloons acquire a slight surface electrostatic charge that results in the adoption of roughly the same geometries when they are tied together at their stems as the corresponding number of electron pairs. For example, five balloons tied together adopt the trigonal bipyramidal geometry, just as do the five bonding pairs of a PCl_5 molecule (AX_5) or the two bonding and three non-bonding pairs of a XeF_2 molecule (AX_2E_3). The molecular geometry of the former is also trigonal bipyramidal, whereas that of the latter is linear.

Examples

The methane molecule (CH_4) is tetrahedral because there are four pairs of electrons. The four hydrogen atoms are positioned at the vertices of a tetrahedron, and the bond angle is $\cos^{-1}(-\frac{1}{3}) \approx 109° 28'$. This is referred to as an AX_4 type of molecule. As mentioned above, A represents the central atom and X represents an outer atom.

The ammonia molecule (NH_3) has three pairs of electrons involved in bonding, but there is a lone pair of electrons on the nitrogen atom. It is not bonded with another atom; however, it influences the overall shape through repulsions. As in methane above, there are four regions of electron density. Therefore, the overall orientation of the regions of electron density is tetrahedral. On the other hand, there are only three outer atoms. This is referred to as an AX_3E type molecule because the lone pair is represented by an E. By definition, the molecular shape or geometry describes the geometric arrangement of the atomic nuclei only, which is trigonal-pyramidal for NH_3.

Steric numbers of 7 or greater are possible, but are less common. The steric number of 7 occurs in iodine heptafluoride (IF_7); the base geometry for a steric number of 7 is pentagonal bipyramidal. The most common geometry for a steric number of 8 is a square antiprismatic geometry. Examples of this include the octacyanomolybdate ($Mo(CN)_8^{4-}$) and octafluorozirconate (ZrF_8^{4-}) anions.

The nonahydridorhenate ion (ReH_9^{2-}) in potassium nonahydridorhenate is a rare example of a compound with a steric number of 9, which has a tricapped trigonal prismatic geometry. Another example is the octafluoroxenate ion (XeF_8^{2-}) in nitrosonium octafluoroxenate(VI), although in this case one of the electron pairs is a lone pair, and therefore the molecule actually has a distorted square antiprismatic geometry.

Possible geometries for steric numbers of 10, 11, 12, or 14 are bicapped square antiprismatic (or bicapped dodecadeltahedral), octadecahedral, icosahedral, and bicapped hexagonal antiprismatic, respectively. No compounds with steric numbers this high involving monodentate ligands exist, and those involving multidentate ligands can often be analysed more simply as complexes with lower steric numbers when some multidentate ligands are treated as a unit.

Exceptions

There are groups of compounds where VSEPR fails to predict the correct geometry.

Some AX_2E_0 Molecules

The gas phase structures of the triatomic halides of the heavier members of group 2, (i.e., calcium, strontium and barium halides, MX_2), are not linear as predicted but are bent, (approximate X–M–X angles: CaF_2, 145°; SrF_2, 120°; BaF_2, 108°; $SrCl_2$, 130°; $BaCl_2$, 115°; $BaBr_2$, 115°; BaI_2, 105°). It has been proposed by Gillespie that this is caused by interaction of the ligands with the electron core of the metal atom, polarising it so that the inner shell is not spherically symmetric, thus influencing the molecular geometry. Ab initio calculations have been cited to propose that contributions from d orbitals in the shell below the valence shell are responsible. Disilynes are also bent, despite having no lone pairs.

Some AX_2E_2 Molecules

One example of the AX_2E_2 geometry is molecular lithium oxide, Li_2O, a linear rather than bent structure, which is ascribed to its bonds being essentially ionic and the strong lithium-lithium repulsion that results. Another example is $O(SiH_3)_2$ with an Si–O–Si angle of 144.1°, which compares to the angles in Cl_2O (110.9°), $(CH_3)_2O$ (111.7°), and $N(CH_3)_3$ (110.9°). Gillespie and Robinson rationalize the Si–O–Si bond angle based on the observed ability of a ligand's lone pair to most greatly repel other electron pairs when the ligand electronegativity is greater than or equal to that of the central atom. In $O(SiH_3)_2$, the central atom is more electronegative, and the lone pairs are less localized and more weakly repulsive. The larger Si–O–Si bond angle results from this and strong ligand-ligand repulsion by the relatively large -SiH_3 ligand. Burford et al showed through X-ray diffraction studies that $Cl_3Al–O–PCl_3$ has a linear Al–O–P bond angle and is therefore a non-VSEPR molecule.

Some AX_6E_1 and AX_8E_1 Molecules\

Some AX_6E_1 molecules, e.g. xenon hexafluoride (XeF_6) and the Te(IV) and Bi(III) anions, $TeCl_6^{2-}$, $TeBr_6^{2-}$, $BiCl_6^{3-}$, $BiBr_6^{3-}$ and BiI_6^{3-}, are octahedra, rather than pentagonal pyramids, and the lone pair does not affect the geometry to the degree predicted by VSEPR. One rationalization is that steric crowding of the ligands allows little or no room for the non-bonding lone pair; another rationalization is the inert pair effect.

Xenon hexafluoride, which has a distorted octahedral geometry.

Transition Metal Molecules

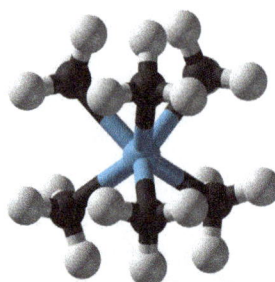

Hexamethyltungsten, a transition metal compound whose
geometry is different from main group coordination.

Many transition metal compounds have unusual geometries, which can be ascribed to ligand bonding interaction with the d subshell and to absence of valence shell lone pairs. Gillespie suggested that this interaction can be weak or strong. Weak interaction is dealt with by the Kepert model, while strong interaction produces bonding pairs that also occupy the respective antipodal points of the sphere. This is similar to predictions based on sd hybrid orbitals using the VALBOND theory. The repulsion of these bidirectional bonding pairs leads to a different prediction of shapes.

Molecule type	Shape	Geometry	Examples
AX_2	Bent		VO_2^+
AX_3	Trigonal pyramidal		CrO_3
AX_4	Tetrahedral		$TiCl_4$

| AX$_5$ | Square pyramidal | 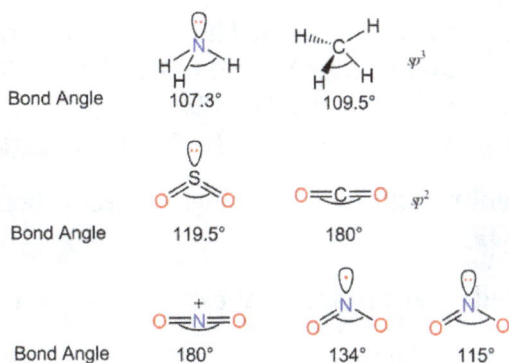 | Ta(CH$_3$)$_5$ |
| AX$_6$ | Trigonal prismatic | | W(CH$_3$)$_6$ |

The square planar shape associated with a d^8 electronic configuration is an exception to the Kepert model. This can be rationalized by considering the increased crystal field stabilization energy as compared to a tetrahedral geometry.

Odd-electron Molecules

The VSEPR theory can be extended to molecules with an odd number of electrons by treating the unpaired electron as a "half electron pair" — for example, Gillespie and Nyholm suggested that the decrease in the bond angle in the series NO_2^+ (180°), NO_2 (134°), NO_2^- (115°) indicates that a given set of bonding electron pairs exert a weaker repulsion on a single non-bonding electron than on a pair of non-bonding electrons. In effect, they considered nitrogen dioxide as an AX$_2$E$_{0.5}$ molecule, with a geometry intermediate between NO_2^+ and NO_2^-. Similarly, chlorine dioxide (ClO_2) is an AX$_2$E$_{1.5}$ molecule, with a geometry intermediate between ClO_2^+ and ClO_2^-.

Finally, the methyl radical (CH$_3$) is predicted to be trigonal pyramidal like the methyl anion (CH_3^-), but with a larger bond angle (as in the trigonal planar methyl cation (CH_3^+)). However, in this case, the VSEPR prediction is not quite true, as CH$_3$ is actually planar, although its distortion to a pyramidal geometry requires very little energy.

Valence Shell Electron Pair Repulsion (VSEPR) Theory:

(i) Lone pairs of electrons (lp) repel each other more strongly than that of bond pair (bp) of electrons. The decreasing order of repulsion is lp - lp > lp - bp > bp - bp.

Examples:

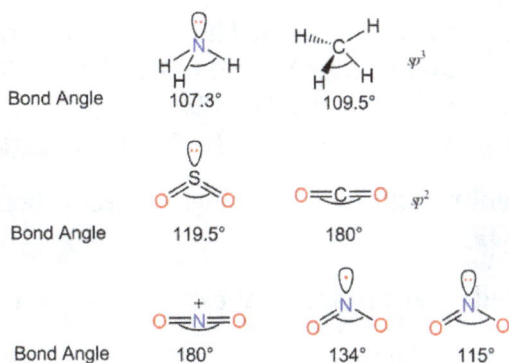

- In CH_4 only bp-bp repulsion exist while, in NH_3 two types of repulsive force present, lp - bp repulsion and bp - bp repulsion. lp - bp repulsion dominates over bp - bp repulsion. Because of this the bonds in NH_3 is more compressed than that in CH_4 which provides lower band angle in NH_3.

- In CO_2 the central C atom does not occupy any lone pair of electron, hence in CO_2 only bp - bp repulsion exists. The central S atom in SO_2 contains one lone pair of electron. Therefore, in SO_2 lp - bp repulsion and bp - bp repulsion present. As we know that lp - bp repulsion dominates over bp - bp repulsion, the O-S-O bonds is compressed by dominating lp - bp.

- NO_2^+ does not contain any paired or unpaired electron on the central N atom. Here, only bp - bp repulsion exists. In NO_2 the central N atom contains one unpaired electron, while in NO_2^- the N atom contains one lone pair of electrons. As in NO_2 only one unpaired electron present, the lp - bp repulsion is lower compared to lp - bp repulsion in NO_2^-. Hence, the dominating lp - bp is more pronounced in NO_2^- compared to NO_2.

(ii) Repulsion between bond pairs decreases as the electronegativity of the atom bound to the central atom increases.

A bond is formed by the sharing of electrons. When the difference in electronegativity between the two connecting atoms increases the electron pair is shifted towards the more electronegative atom. With increases of electronegativity this shift is more to the electronegative atom. This decreases the repulsion effect between two bond pairs.

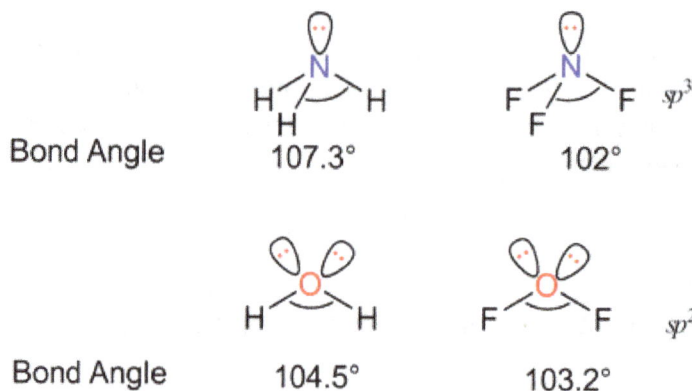

Bond Angle	$107.3°$	$102°$
Bond Angle	$104.5°$	$103.2°$

- F is more electronegative that that of H. Therefore, compared to NH_3, in NF_3 the bond forming lone pair of electron is more shifted to the F atom. This decreases bp - bp repulsion. Therefore, in NF_3 the lp - bp repulsion is more predominating that that of NH_3 and makes F-N-F bond angle $102°$ compared to H-N-H bond angle $107.3°$.

- Same (above) argument is valid for having higher H-O-H bond angle in H_2O compared to F-O-F bond angle in OF_2.

(iii) Electronpairs in filled shell repel stronger that electronpairs in incomplete shell.

When the electron pair containing atom contains low-energy lying suitable vacant orbitals the electron pair of electrons can undergoes diffusion to the vacant orbitals. Hence, the lp - bp repulsion diminishes dramatically and bp - bp repulsion dominates. In this case, the bond angles around the central metal increase.

Example:

Bond Angle	107.3°	93.8°

Bond Angle	104.5°	92.2°

Bond Angle	91°	89.5°

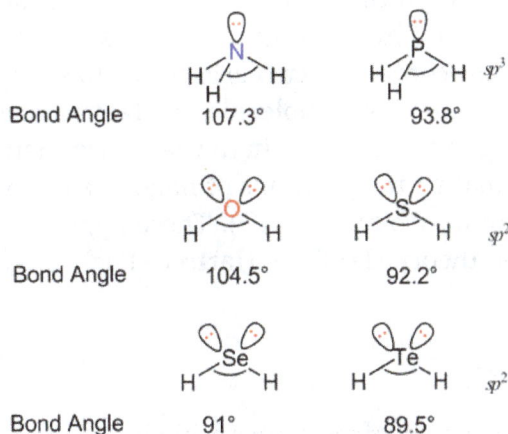

- On going from O to Te the size of the central atom increases. Therefore, diffusion is more predominating which further decreases lp - bp repulsion.

(IV) Lone pair of electron can be transferred from a filled shell to a energetically suitable empty shell of the other bonded atom.

Due to transfer of lone pair of electron the bond accumulate some sort of double bond character. This enhances the bp - bp repulsion and consequently, causing a bigger bond angle.

Example:

Bond Angle	93.8°	97.8°

Bond Angle	104.5°	103.2°	110.8°

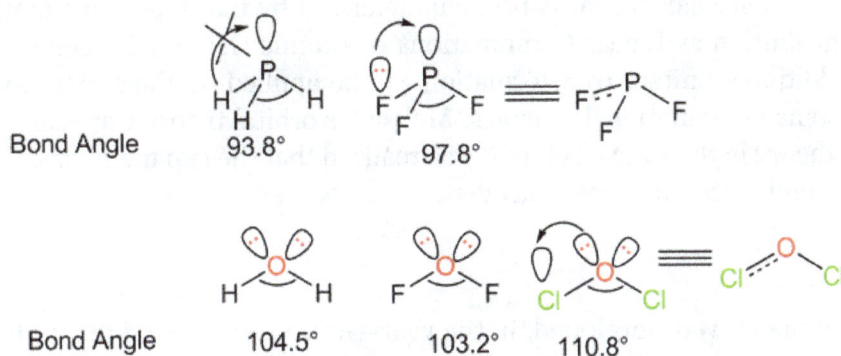

- P and Cl contain low-lying vacant d orbitals. Lone pair of electron from the filled orbitals of F or O can be transferred to the vacant orbital. This causes a partial - multiple bond that enhances the bp - bp repulsion. Hence, the bond angle increases.

(V) Multiple bond orbitals repel each other more strongly than single bond orbitals.

Examples:

$O = PF_3$ (F-P-F bond angle = 103°), while in $O = PBr_3$ (Br-P-Br bond angle = 108°)

Molecular Orbital Theory

In chemistry, molecular orbital (MO) theory is a method for determining molecular structure in which electrons are not assigned to individual bonds between atoms, but are treated as moving under the influence of the nuclei in the whole molecule. The spatial and energetic properties of electrons within atoms are fixed by quantum mechanics to form orbitals that contain these electrons. While atomic orbitals contain electrons ascribed to a single atom, molecular orbitals, which surround a number of atoms in a molecule, contain valence electrons between atoms. Molecular orbital theory, which was proposed in the early twentieth century, revolutionized the study of bonding by approximating the positions of bonded electrons—the molecular orbitals—as linear combinations of atomic orbitals (LCAO). These approximations are now made by applying the density functional theory (DFT) or Hartree–Fock (HF) models to the Schrödinger equation.

Quantitative Applications

In this theory, each molecule has a set of molecular orbitals, in which it is assumed that the molecular orbital wave function ψ_j can be written as a simple weighted sum of the n constituent atomic orbitals χ_i, according to the following equation:

$$\psi_j = \sum_{i=1}^{n} c_{ij} \chi_i.$$

One may determine c_{ij} coefficients numerically by substituting this equation into the Schrödinger's equation and applying the variational principle. The variational principle is a mathematical technique used in quantum mechanics to build up the coefficients of each atomic orbital basis. A larger coefficient means that the orbital basis is composed more of that particular contributing atomic orbital—hence, the molecular orbital is best characterized by that type. This method of quantifying orbital contribution as Linear Combinations of Atomic Orbitals is used in computational chemistry. An additional unitary transformation can be applied on the system to accelerate the convergence in some computational schemes. Molecular orbital theory was seen as a competitor to valence bond theory in the 1930s, before it was realized that the two methods are closely related and that when extended they become equivalent.

History

Molecular orbital theory was developed, in the years after valence bond theory had been established (1927), primarily through the efforts of Friedrich Hund, Robert Mulliken, John C. Slater, and John Lennard-Jones. MO theory was originally called the Hund-Mulliken theory. According to German physicist and physical chemist Erich Hückel, the first quantitative use of molecular orbital theory was the 1929 paper of Lennard-Jones. This paper notably predicted a triplet ground state for the dioxygen molecule which explained its paramagnetism before valence bond theory, which came up with its own explanation in 1931. The word *orbital* was introduced by Mulliken in 1932. By 1933, the molecular orbital theory had been accepted as a valid and useful theory.

Erich Hückel applied molecular orbital theory to unsaturated hydrocarbon molecules starting in 1931 with his Hückel molecular orbital (HMO) method for the determination of MO energies for pi electrons, which he applied to conjugated and aromatic hydrocarbons. This method provided an explanation of the stability of molecules with six pi-electrons such as benzene.

The first accurate calculation of a molecular orbital wavefunction was that made by Charles Coulson in 1938 on the hydrogen molecule. By 1950, molecular orbitals were completely defined as eigenfunctions (wave functions) of the self-consistent field Hamiltonian and it was at this point that molecular orbital theory became fully rigorous and consistent. This rigorous approach is known as the Hartree–Fock method for molecules although it had its origins in calculations on atoms. In calculations on molecules, the molecular orbitals are expanded in terms of an atomic orbital basis set, leading to the Roothaan equations. This led to the development of many ab initio quantum chemistry methods. In parallel, molecular orbital theory was applied in a more approximate manner using some empirically derived parameters in methods now known as semi-empirical quantum chemistry methods.

The success of Molecular Orbital Theory also spawned ligand field theory, which was developed during the 1930s and 1940s as an alternative to crystal field theory.

Types of Orbitals

Molecular orbital (MO) theory uses a linear combination of atomic orbitals (LCAO) to represent molecular orbitals resulting from bonds between atoms. These are often divided into *bonding* orbitals, anti-bonding orbitals, and non-bonding orbitals. A bonding orbital concentrates electron density in the region *between* a given pair of atoms, so that its electron density will tend to attract each of the two nuclei toward the other and hold the two atoms together. An anti-bonding orbital concentrates electron density "behind" each nucleus (i.e. on the side of each atom which is farthest from the other atom), and so tends to pull each of the two nuclei away from the other and actually weaken the bond between the two nuclei. Electrons in non-bonding orbitals tend to be associated with atomic orbitals that do not interact positively or negatively with one another, and electrons in these orbitals neither contribute to nor detract from bond strength.

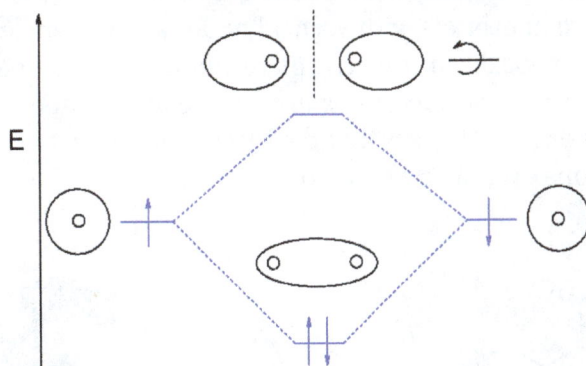

MO diagram showing the formation of molecular orbitals of H_2 (centre) from atomic orbitals of two H atoms. The lower-energy MO is bonding with electron density concentrated between the two H nuclei. The higher-energy MO is anti-bonding with electron density concentrated behind each H nucleus.

Molecular orbitals are further divided according to the types of atomic orbitals they are formed from. Chemical substances will form bonding interactions if their orbitals become lower in energy when they interact with each other. Different bonding orbitals are distinguished that differ by electron configuration (electron cloud shape) and by energy levels.

The molecular orbitals of a molecule can be illustrated in molecular orbital diagrams.

Overview

MO theory provides a global, delocalized perspective on chemical bonding. In MO theory, *any* electron in a molecule may be found *anywhere* in the molecule, since quantum conditions allow electrons to travel under the influence of an arbitrarily large number of nuclei, as long as they are in eigenstates permitted by certain quantum rules. Thus, when excited with the requisite amount of energy through high-frequency light or other means, electrons can transition to higher-energy molecular orbitals. For instance, in the simple case of a hydrogen diatomic molecule, promotion of a single electron from a bonding orbital to an antibonding orbital can occur under UV radiation. This promotion weakens the bond between the two hydrogen atoms and can lead to photodissociation—the breaking of a chemical bond due to the absorption of light.

Although in MO theory *some* molecular orbitals may hold electrons that are more localized between specific pairs of molecular atoms, *other* orbitals may hold electrons that are spread more uniformly over the molecule. Thus, overall, bonding is far more delocalized in MO theory, which makes it more applicable to resonant molecules that have equivalent non-integer bond orders than valence bond (VB) theory. This makes MO theory more useful for the description of extended systems.

An example is the MO description of benzene, C_6H_6, which is an aromatic hexagonal ring of six carbon atoms and three double bonds. In this molecule, 24 of the 30 total valence bonding electrons—24 coming from carbon atoms and 6 coming from hydrogen atoms—are located in 12 σ (sigma) bonding orbitals, which are located mostly between pairs of atoms (C-C or C-H), similarly to the electrons in the valence bond description. However, in benzene the remaining six bonding electrons are located in three π (pi) molecular bonding orbitals that are delocalized around the ring. Two of these electrons are in an MO that has equal orbital contributions from all six atoms. The other four electrons are in orbitals with vertical nodes at right angles to each other. As in the VB theory, all of these six delocalized π electrons reside in a larger space that exists above and below the ring plane. All carbon-carbon bonds in benzene are chemically equivalent. In MO theory this is a direct consequence of the fact that the three molecular π orbitals combine and evenly spread the extra six electrons over six carbon atoms.

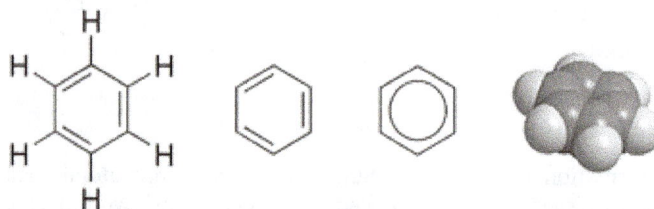

Structure of benzene

In molecules such as methane, CH_4, the eight valence electrons are found in four MOs that are spread out over all five atoms. However, it is possible to transform the MOs into four localized sp³ orbitals. Linus Pauling, in 1931, hybridized the carbon 2s and 2p orbitals so that they pointed directly at the hydrogen 1s basis functions and featured maximal overlap. However, the delocalized MO description is more appropriate for predicting ionization energies and the positions of spectral absorption bands. When methane is ionized, a single electron is taken from the valence MOs, which can come from the s bonding or the triply degenerate p bonding levels, yielding two ionization energies. In comparison, the explanation in VB theory is more complicated. When one electron is removed from an sp³ orbital, resonance is invoked between four valence bond structures, each of which has a single one-electron bond and three two-electron bonds. Triply degenerate T_2 and A_1 ionized states (CH_4^+) are produced from different linear combinations of these four structures. The difference in energy between the ionized and ground state gives the two ionization energies.

As in benzene, in substances such as beta carotene, chlorophyll, or heme, some electrons in the π orbitals are spread out in molecular orbitals over long distances in a molecule, resulting in light absorption in lower energies (the visible spectrum), which accounts for the characteristic colours of these substances. This and other spectroscopic data for molecules are well explained in MO theory, with an emphasis on electronic states associated with multicenter orbitals, including mixing of orbitals premised on principles of orbital symmetry matching. The same MO principles also naturally explain some electrical phenomena, such as high electrical conductivity in the planar direction of the hexagonal atomic sheets that exist in graphite. This results from continuous band overlap of half-filled p orbitals and explains electrical conduction. MO theory recognizes that some electrons in the graphite atomic sheets are completely delocalized over arbitrary distances, and reside in very large molecular orbitals that cover an entire graphite sheet, and some electrons are thus as free to move and therefore conduct electricity in the sheet plane, as if they resided in a metal.

In the Valence bond theory, the individuality of atomic properties of the constituting atoms forming a molecule is retained too some extent. Whereas, in the molecular orbital theory it is considered that molecular orbitals are formed by the linear combination of the two constituting atomic orbitals´ wave functions.

Let us consider, $\psi(A)$ and $\psi(B)$ are two atomic orbitals which are combining to form molecular orbitals for A - B bond. Molecular orbitals will also be wave functions anf they can be represented as;

$\psi(MO_b) = \psi(A) + \psi(B)$; y (MO_b) = bonding MO wave function

and

$\psi(MO_{ab}) = \psi(A) - \psi(B)$; $\psi(MO_{ab})$ = anti-bonding MO wave function

Conditions for the favorable combination of atomic orbitals:

(A) Atomic orbitals must have comparable energies.

(B) They must have comparable energies.

(C) They must have same symmetry with respect to the bonding molecular axis, i.e. if the atomic orbitals are perpendicular to each other there will be no overlap or linear combination between them.

Bond order:

Bond order (BO) of a bond is defined as;

BO = [No of electron in bonding MO - no of electron in anti - bonding MO]/2

Starting from hydrogen to fluorine atom the energy difference between the 2 s and 2 p orbitals increases due to increase in electronegativity. It is important to note that up to nitrogen, the energy difference between the 2s and 2p orbital is sufficient for sp type hybridization. Therefore, up to nitrogen nitrogen molecule the energy level follow the order 1σ , 1σ * , 2π , 2σ , 2σ * , 2π * (i.e. energy of 2σ *> 2π*) and for oxygen and nitrogen molecules the order is 1σ , 1σ * , 2σ , 2π , 2π * , 2σ *are (i.e. energy of 2σ *<2π *).

MOs of H_2 Molecule:

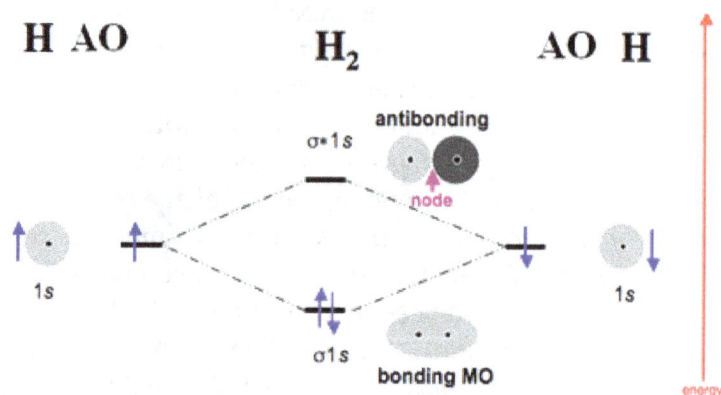

Bond Order = [2-0]/2 = 2/2 = 1

Here * represents anti-bonding orbital.

MOs of He_2 Molecule:

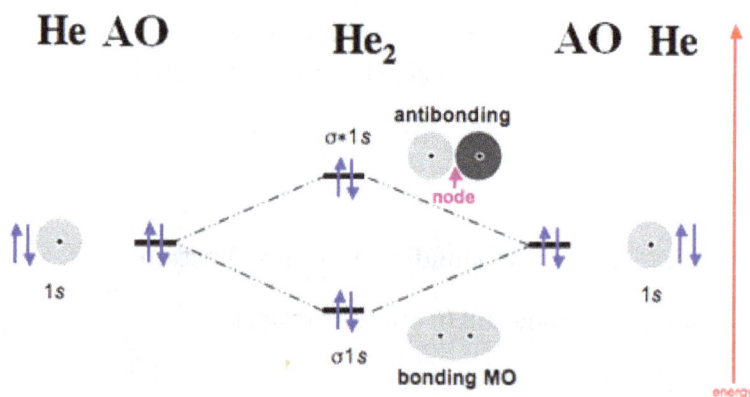

Bond Order = [2-2]/2 = 0/2 = 0.

As the BO is zero, there is no bond between He_2 molecule. This means He is monoatomic molecule.

MOs of Li_2 Molecule:

Li_2

AOs Li   AOs Li

$\sigma_{2s}{}^{*}$

$2s$ $2s$

σ_{2s}

$\sigma_{1s}{}^{*}$

$1s$ $1s$

σ_{1s}

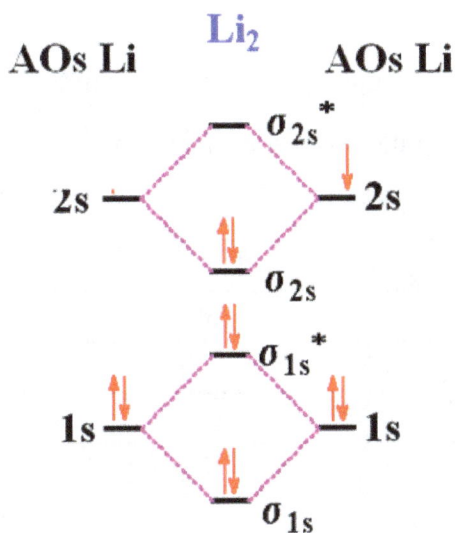

Bond Order = [4-2]/2 = 2/2 = 1

MOs of Be_2 Molecule:

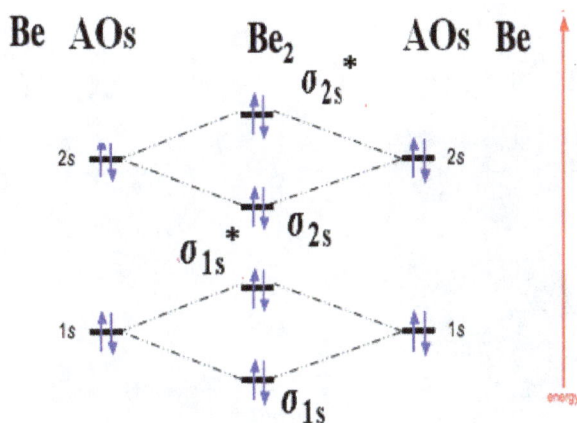

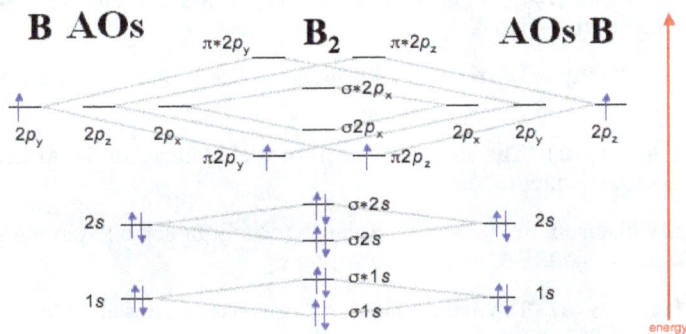

Be AOs Be_2 AOs Be

$\sigma_{2s}{}^{*}$

$2s$ $2s$

σ_{2s}

$\sigma_{1s}{}^{*}$

$1s$ $1s$

σ_{1s}

energy

Bond Order = [4-4]/2 = 0/2 = 0

That is no bond between Be atom. Be molecule is monoatomic.

MOs of B_2 Molecule:

B AOs B_2 AOs B

$\pi^{*}2p_y$ $\pi^{*}2p_z$

$\sigma^{*}2p_x$

$2p_y$ $2p_z$ $2p_x$ $\sigma 2p_x$ $2p_x$ $2p_y$ $2p_z$

$\pi 2p_y$ $\pi 2p_z$

$\sigma^{*}2s$

$2s$ $2s$

$\sigma 2s$

$\sigma^{*}1s$

$1s$ $1s$

$\sigma 1s$

energy

Bond Order = [6-4]/2 = 2/2 = 1

Carbon and nitrogen have same MO energy distribution as in B_2 .

In Carbon and nitrogen the electron distributions in MO are;

	Carbon	Nitrogen
σ1S	2	2
σ*1S	2	2
σ2S	2	2
σ*2S	2	2
π2p$_y$	2	2
π2pz	2	2
σ2P$_x$	0	2
Bond order	[8-4]/2 = 4/2 = 2	[10-4]/2 = 6/2 = 3

MOs of B_2 Molecule:

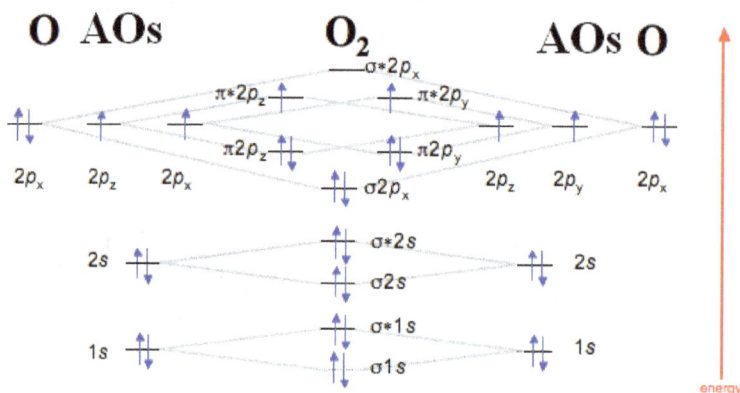

References

- Lewis, Gilbert N. (1916). "The Atom and the Molecule". Journal of the American Chemical Society. 38 (4): 772. doi:10.1021/ja02261a002

- Atkins, Peter; Loretta Jones (1997). Chemistry: Molecules, Matter and Change. New York: W. H. Freeman & Co. pp. 294–295. ISBN 0-7167-3107-X

- Gillespie, R. J. (2008). "Fifty years of the VSEPR model". Coord. Chem. Rev. 252: 1315–1327. doi:10.1016/j. ccr.2007.07.007

- Lewis, Gilbert N. (1916-04-01). "The atom and the molecule". Journal of the American Chemical Society. 38 (4): 762–785. doi:10.1021/ja02261a002

- Campbell, Neil A.; Williamson, Brad; Heyden, Robin J. (2006). Biology: Exploring Life. Boston, MA: Pearson Prentice Hall. ISBN 0-13-250882-6. Retrieved 2012-02-05

- Langmuir, Irving (1919-06-01). "The Arrangement of Electrons in Atoms and Molecules". Journal of the American Chemical Society. 41 (6): 868–934. doi:10.1021/ja02227a002

- Brittin, W. E. (1945). "Valence Angle of the Tetrahedral Carbon Atom". J. Chem. Educ. 22 (3): 145. doi:10.1021/ed022p145

- Hanson, Robert M. (1995). Molecular origami: precision scale models from paper. University Science Books. ISBN 0-935702-30-X

- James, H. H.; Coolidge, A. S. (1933). "The Ground State of the Hydrogen Molecule". Journal of Chemical Physics. American Institute of Physics. 1 (12): 825–835. doi:10.1063/1.1749252

- Anslyn, E. V.; Dougherty, D. A. (2006). Modern Physical Organic Chemistry. University Science Books. p. 57. ISBN 978-1891389313

- Hall, G.G. (7 August 1950). "The Molecular Orbital Theory of Chemical Valency. VI. Properties of Equivalent Orbitals" (pdf). Proc. Roy. Soc. A. 202 (1070): 336–344. Bibcode:1950RSPSA.202..336H. doi:10.1098/rspa.1950.0104

Acids-Bases and its Related Theories

A chemical reaction occurring between a base and an acid is known as an acid-base reaction. Theories that provide conceptual frameworks to the reaction mechanisms of acid-base reactions include the Arrhenius theory, the Brønsted–Lowry acid–base theory and the Franklin theory. The topics discussed in the chapter are of great importance to broaden the existing knowledge on inorganic chemistry.

Acid–base Reaction

An acid–base reaction is a chemical reaction that occurs between an acid and a base. Several theoretical frameworks provide alternative conceptions of the reaction mechanisms and their application in solving related problems; these are called acid–base theories, for example, Brønsted–Lowry acid–base theory. Their importance becomes apparent in analyzing acid–base reactions for gaseous or liquid species, or when acid or base character may be somewhat less apparent. The first of these concepts was provided by the French chemist Antoine Lavoisier, around 1776.

Historic Development

Lavoisier's Oxygen Theory of Acids

The first scientific concept of acids and bases was provided by Lavoisier in around 1776. Since Lavoisier's knowledge of strong acids was mainly restricted to oxoacids, such as HNO_3 (nitric acid) and H_2SO_4 (sulfuric acid), which tend to contain central atoms in high oxidation states surrounded by oxygen, and since he was not aware of the true composition of the hydrohalic acids (HF, HCl, HBr, and HI), he defined acids in terms of their containing *oxygen*, which in fact he named from Greek words meaning "acid-former". The Lavoisier definition was held as absolute truth for over 30 years, until the 1810 article and subsequent lectures by Sir Humphry Davy in which he proved the lack of oxygen in H_2S, H_2Te, and the hydrohalic acids. However, Davy failed to develop a new theory, concluding that "acidity does not depend upon any particular elementary substance, but upon peculiar arrangement of various substances". One notable modification of oxygen theory was provided by Berzelius, who stated that acids are oxides of nonmetals while bases are oxides of metals.

Liebig's Hydrogen Theory of Acids

In 1838, Justus von Liebig proposed that an acid is a hydrogen-containing substance in which the hydrogen could be replaced by a metal. This redefinition was based on his extensive work on the chemical composition of organic acids, finishing the doctrinal shift from oxygen-based acids to hydrogen-based acids started by Davy. Liebig's definition, while completely empirical, remained in use for almost 50 years until the adoption of the Arrhenius definition.

Arrhenius Definition

Svante Arrhenius

The first modern definition of acids and bases in molecular terms was devised by Svante Arrhenius. A hydrogen theory of acids, it followed from his 1884 work with Friedrich Wilhelm Ostwald in establishing the presence of ions in aqueous solution and led to Arrhenius receiving the Nobel Prize in Chemistry in 1903.

As defined by Arrhenius:

- *an Arrhenius acid* is a substance that dissociates in water to form hydrogen ions (H^+); that is, an acid increases the concentration of H^+ ions in an aqueous solution.

This causes the protonation of water, or the creation of the hydronium (H_3O^+) ion. Thus, in modern times, the symbol H^+ is interpreted as a shorthand for H_3O^+, because it is now known that a bare proton does not exist as a free species in aqueous solution.

- *an Arrhenius base* is a substance that dissociates in water to form hydroxide (OH^-) ions; that is, a base increases the concentration of OH^- ions in an aqueous solution.

The Arrhenius definitions of acidity and alkalinity are restricted to aqueous solutions, and refer to the concentration of the solvent ions. Under this definition, pure H_2SO_4 and HCl dissolved in toluene are not acidic, and molten NaOH and solutions of calcium amide in liquid ammonia are not alkaline.

Overall, to qualify as an Arrhenius acid, upon the introduction to water, the chemical must either cause, directly or otherwise:

- an increase in the aqueous hydronium concentration, or

- a decrease in the aqueous hydroxide concentration.

Conversely, to qualify as an Arrhenius base, upon the introduction to water, the chemical must either cause, directly or otherwise:

- a decrease in the aqueous hydronium concentration, or

- an increase in the aqueous hydroxide concentration.

The reaction of an acid with a base is called a neutralization reaction. The products of this reaction are a salt and water.

$$acid + base \rightarrow salt + water$$

In this traditional representation an acid–base neutralization reaction is formulated as a double-replacement reaction. For example, the reaction of hydrochloric acid, HCl, with sodium hydroxide, NaOH, solutions produces a solution of sodium chloride, NaCl, and some additional water molecules.

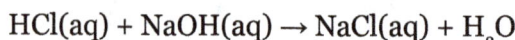

$$HCl(aq) + NaOH(aq) \rightarrow NaCl(aq) + H_2O$$

The modifier (aq) in this equation is important. It was implied by Arrhenius, not included explicitly. It indicates that the substances are dissolved in water. In fact though all three substances, HCl, NaOH and NaCl are capable of existing as pure compounds, in aqueous solutions they are fully dissociated into the (aquated) ions H^+, Cl^-, Na^+ and OH^-.

Brønsted–Lowry Definition

Johannes Nicolaus Brønsted and Thomas Martin Lowry

The Brønsted–Lowry definition, formulated in 1923, independently by Johannes Nicolaus Brønsted in Denmark and Martin Lowry in England, is based upon the idea of protonation of bases through the de-protonation of acids – that is, the ability of acids to "donate" hydrogen ions (H^+)—otherwise known as protons—to bases, which "accept" them.

An acid–base reaction is, thus, the removal of a hydrogen ion from the acid and its addition to the base. The removal of a hydrogen ion from an acid produces its *conjugate base*, which is the acid with a hydrogen ion removed. The reception of a proton by a base produces its *conjugate acid*, which is the base with a hydrogen ion added.

Unlike the previous definitions, the Brønsted–Lowry definition does not refer to the formation of salt and solvent, but instead to the formation of *conjugate acids* and *conjugate bases*, produced by the transfer of a proton from the acid to the base. In this approach, acids and bases are fundamentally different in behavior from salts, which are seen as electrolytes, subject to the theories of Debye, Onsager, and others. An acid and a base react not to produce a salt and a solvent, but to form a new acid and a new base. The concept of neutralization is thus absent. Brønsted–Lowry acid–base behavior is formally independent of any solvent, making it more all-encompassing than the Arrhenius model.

The general formula for acid–base reactions according to the Brønsted–Lowry definition is:

$$HA + B \rightarrow BH^+ + A^-$$

where HA represents the acid, B represents the base, BH^+ represents the conjugate acid of B, and A^- represents the conjugate base of HA.

For example, a Brønsted-Lowry model for the dissociation of hydrochloric acid (HCl) in aqueous solution would be the following:

$$HCl + H_2O \rightleftharpoons H_3O^+ + Cl^-$$

The removal of H^+ from the HCl produces the chloride ion, Cl^-, the conjugate base of the acid. The addition of H^+ to the H_2O (acting as a base) forms the hydronium ion, H_3O^+, the conjugate acid of the base.

Water is amphoteric—that is, it can act as both an acid and a base. The Brønsted-Lowry model explains this, showing the dissociation of water into low concentrations of hydronium and hydroxide ions:

$$H_2O + H_2O \rightleftharpoons H_3O^+ + OH^-$$

This equation is demonstrated in the image below:

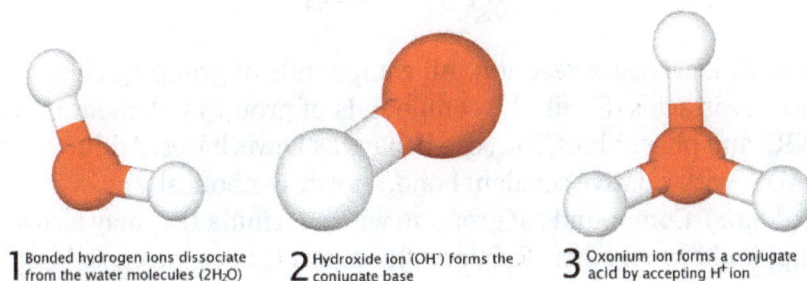

1 Bonded hydrogen ions dissociate from the water molecules ($2H_2O$)

2 Hydroxide ion (OH^-) forms the conjugate base

3 Oxonium ion forms a conjugate acid by accepting H^+ ion

Here, one molecule of water acts as an acid, donating an H^+ and forming the conjugate base, OH^-, and a second molecule of water acts as a base, accepting the H^+ ion and forming the conjugate acid, H_3O^+.

As an example of water acting as an acid, consider an aqueous solution of pyridine, C_5H_5N.

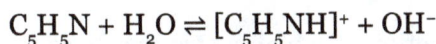

$$C_5H_5N + H_2O \rightleftharpoons [C_5H_5NH]^+ + OH^-$$

In this example, a water molecule is split into a hydrogen ion, which is donated to a pyridine molecule, and an hydroxide ion.

In the Brønsted-Lowry model, the solvent does not necessarily have to be water. For example, consider what happens when acetic acid, CH_3COOH, dissolves in liquid ammonia.

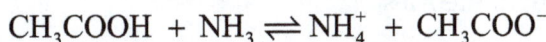

$$CH_3COOH + NH_3 \rightleftharpoons NH_4^+ + CH_3COO^-$$

An H^+ ion is removed from acetic acid, forming its conjugate base, the acetate ion, CH_3COO^-. The addition of an H^+ ion to an ammonia molecule of the solvent creates its conjugate acid, the ammonium ion, NH_4^+.

The Brønsted–Lowry model calls hydrogen-containing substances (like HCl) acids. Thus, some substances, which many chemists considered to be acids, such as SO_3 or BCl_3, are excluded from this classification due to lack of hydrogen. Gilbert N. Lewis wrote in 1938, "To restrict the group of acids to those substances that contain hydrogen interferes as seriously with the systematic understanding of chemistry as would the restriction of the term oxidizing agent to substances containing oxygen." Furthermore, KOH and KNH_2 are not considered Brønsted bases, but rather salts containing the bases OH^- and NH_2^-.

Lewis Definition

The hydrogen requirement of Arrhenius and Brønsted–Lowry was removed by the Lewis definition of acid–base reactions, devised by Gilbert N. Lewis in 1923, in the same year as Brønsted–Lowry, but it was not elaborated by him until 1938. Instead of defining acid–base reactions in terms of protons or other bonded substances, the Lewis definition defines a base (referred to as a *Lewis base*) to be a compound that can donate an *electron pair*, and an acid (a *Lewis acid*) to be a compound that can receive this electron pair.

For example, boron trifluoride, BF_3 is a typical Lewis acid. It can accept a pair of electrons as it has a vacancy in its octet. The fluoride ion has a full octet and can donate a pair of electrons. Thus

$$BF_3 + F^- \rightarrow BF_4^-$$

is a typical Lewis acid, Lewis base reaction. All compounds of group 13 elements with a formula AX_3 can behave as Lewis acids. Similarly, compounds of group 15 elements with a formula DY_3, such as amines, NR_3, and phosphines, PR_3, can behave as Lewis bases. Adducts between them have the formula $X_3A \leftarrow DY_3$ with a dative covalent bond, shown symbolically as $\leftarrow$, between the atoms A (acceptor) and D (donor). Compounds of group 16 with a formula DX_2 may also act as Lewis bases; in this way, a compound like an ether, R_2O, or a thioether, R_2S, can act as a Lewis base. The Lewis definition is not limited to these examples. For instance, carbon monoxide acts as a Lewis base when it forms an adduct with boron trifluoride, of formula $F_3B \leftarrow CO$.

Adducts involving metal ions are referred to as co-ordination compounds; each ligand donates a pair of electrons to the metal ion. The reaction

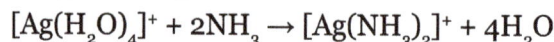

$$[Ag(H_2O)_4]^+ + 2NH_3 \rightarrow [Ag(NH_3)_2]^+ + 4H_2O$$

can be seen as an acid–base reaction in which a stronger base (ammonia) replaces a weaker one (water)

The Lewis and Brønsted–Lowry definitions are consistent with each other since the reaction

$$H^+ + OH^- \rightleftharpoons H_2O$$

is an acid–base reaction in both theories.

Solvent System Definition

One of the limitations of the Arrhenius definition is its reliance on water solutions. Edward Curtis Franklin studied the acid–base reactions in liquid ammonia in 1905 and pointed out the similarities to the water-based Arrhenius theory. Albert F. O. Germann, working with liquid phosgene, $COCl_2$, formulated the solvent-based theory in 1925, thereby generalizing the Arrhenius definition to cover aprotic solvents.

Germann pointed out that in many solutions, there are ions in equilibrium with the neutral solvent molecules:

- solvonium: A generic name for a positive ion.

- solvate: A generic name for a negative ion.

For example, water and ammonia undergo such dissociation into hydronium and hydroxide, and ammonium and amide, respectively:

$$2\,H_2O \rightleftharpoons H_3O^+ + OH^-$$

$$2\,NH_3 \rightleftharpoons NH_4^+ + NH_2^-$$

Some aprotic systems also undergo such dissociation, such as dinitrogen tetroxide into nitrosonium and nitrate, antimony trichloride into dichloroantimonium and tetrachloroantimonate, and phosgene into chlorocarboxonium and chloride:

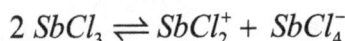

$$N_2O_4 \rightleftharpoons NO^+ + NO_3^-$$

$$2\,SbCl_3 \rightleftharpoons SbCl_2^+ + SbCl_4^-$$

$$COCl_2 \rightleftharpoons COCl^+ + Cl^-$$

A solute that causes an increase in the concentration of the solvonium ions and a decrease in the concentration of solvate ions is defined as an *acid*. A solute that causes an increase in the concentration of the solvate ions and a decrease in the concentration of the solvonium ions is defined as a *base*.

Thus, in liquid ammonia, KNH_2 (supplying NH_2^-) is a strong base, and NH_4NO_3 (supplying NH_4^+) is a strong acid. In liquid sulfur dioxide (SO_2), thionyl compounds (supplying SO^{2+}) behave as acids, and sulfites (supplying SO_3^{2-}) behave as bases.

The non-aqueous acid–base reactions in liquid ammonia are similar to the reactions in water:

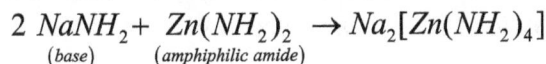

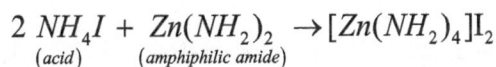

$$\underset{(base)}{2\,NaNH_2} + \underset{(amphiphilic\ amide)}{Zn(NH_2)_2} \rightarrow Na_2[Zn(NH_2)_4]$$

$$\underset{(acid)}{2\,NH_4I} + \underset{(amphiphilic\ amide)}{Zn(NH_2)_2} \rightarrow [Zn(NH_2)_4]I_2$$

Nitric acid can be a base in liquid sulfuric acid:

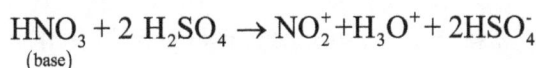

$$\underset{(base)}{HNO_3} + 2\,H_2SO_4 \rightarrow NO_2^+ + H_3O^+ + 2HSO_4^-$$

The unique strength of this definition shows in describing the reactions in aprotic solvents; for example, in liquid N_2O_4:

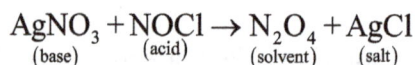

$$\underset{\text{(base)}}{AgNO_3} + \underset{\text{(acid)}}{NOCl} \rightarrow \underset{\text{(solvent)}}{N_2O_4} + \underset{\text{(salt)}}{AgCl}$$

Because the solvent system definition depends on the solute as well as on the solvent itself, a particular solute can be either an acid or a base depending on the choice of the solvent: $HClO_4$ is a strong acid in water, a weak acid in acetic acid, and a weak base in fluorosulfonic acid; this characteristic of the theory has been seen as both a strength and a weakness, because some substances (such as SO_3 and NH_3) have been seen to be acidic or basic on their own right. On the other hand, solvent system theory has been criticized as being too general to be useful. Also, it has been thought that there is something intrinsically acidic about hydrogen compounds, a property not shared by non-hydrogenic solvonium salts.

Lux–Flood Definition

This acid–base theory was a revival of oxygen theory of acids and bases, proposed by German chemist Hermann Lux in 1939, further improved by Håkon Flood circa 1947 and is still used in modern geochemistry and electrochemistry of molten salts. This definition describes an acid as an oxide ion (O^{2-}) acceptor and a base as an oxide ion donor. For example:

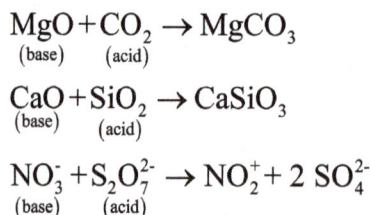

$$\underset{\text{(base)}}{MgO} + \underset{\text{(acid)}}{CO_2} \rightarrow MgCO_3$$

$$\underset{\text{(base)}}{CaO} + \underset{\text{(acid)}}{SiO_2} \rightarrow CaSiO_3$$

$$\underset{\text{(base)}}{NO_3^-} + \underset{\text{(acid)}}{S_2O_7^{2-}} \rightarrow NO_2^+ + 2\,SO_4^{2-}$$

This theory is also useful in the systematisation of the reactions of noble gas compounds, especially the xenon oxides, fluorides, and oxofluorides.

Usanovich Definition

Mikhail Usanovich developed a general theory that does not restrict acidity to hydrogen-containing compounds, but his approach, published in 1938, was even more general than Lewis theory. Usanovich's theory can be summarized as defining an acid as anything that accepts negative species or donates positive ones, and a base as the reverse. This defined the concept of redox (oxidation-reduction) as a special case of acid–base reactions

Some examples of Usanovich acid–base reactions include:

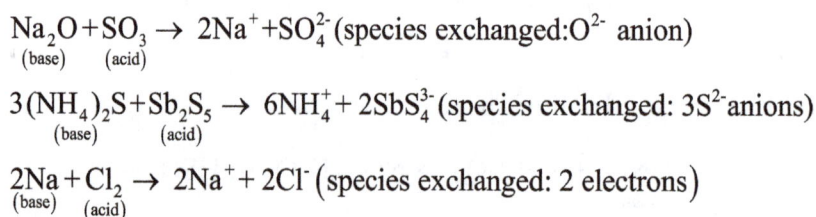

$$\underset{\text{(base)}}{Na_2O} + \underset{\text{(acid)}}{SO_3} \rightarrow 2Na^+ + SO_4^{2-} \text{ (species exchanged:} O^{2-} \text{ anion)}$$

$$\underset{\text{(base)}}{3(NH_4)_2S} + \underset{\text{(acid)}}{Sb_2S_5} \rightarrow 6NH_4^+ + 2SbS_4^{3-} \text{ (species exchanged: } 3S^{2-} \text{anions)}$$

$$\underset{\text{(base)}}{2Na} + \underset{\text{(acid)}}{Cl_2} \rightarrow 2Na^+ + 2Cl^- \text{ (species exchanged: 2 electrons)}$$

Pearson Definition

In 1963, Ralph Pearson proposed a qualitative concept known as Hard Soft Acid Base principle. later made quantitative with help of Robert Parr in 1984. 'Hard' applies to species that are small, have high charge states, and are weakly polarizable. 'Soft' applies to species that are large, have low charge states and are strongly polarizable. Acids and bases interact, and the most stable interactions are hard–hard and soft–soft. This theory has found use in organic and inorganic chemistry.

Acid–base Equilibrium

The reaction of a strong acid with a strong base is essentially a quantitative reaction. For example,

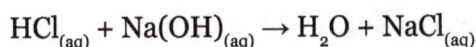

$$HCl_{(aq)} + Na(OH)_{(aq)} \rightarrow H_2O + NaCl_{(aq)}$$

In this reaction both the sodium and chloride ions are spectators as the neutralization reaction,

$$H^+ + OH^- \rightarrow H_2O$$

does not involve them. With weak bases addition of acid is not quantitative because a solution of a weak base is a buffer solution. A solution of a weak acid is also a buffer solution. When a weak acid reacts with a weak base an equilibrium mixture is produced. For example, adenine, written as AH can react with a hydrogen phosphate ion, HPO_4^{2-}.

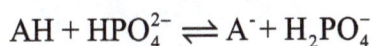

$$AH + HPO_4^{2-} \rightleftharpoons A^- + H_2PO_4^-$$

The equilibrium constant for this reaction can be derived from the acid dissociation constants of adenine and of the dihydrogen phosphate ion.

$$[A^-][H^+] = K_{a1}[AH]$$
$$[HPO_4^{2-}][H^+] = K_{a2}[H_2PO_4^-]$$

The notation [X] signifies "concentration of X". When these two equations are combined by eliminating the hydrogen ion concentration, an expression for the equilibrium constant, K is obtained.

$$[A^-][H_2PO_4^-] = K[AH][HPO_4^{2-}]; \quad K = \frac{K_{a1}}{K_{a2}}$$

Acid–alkali Reaction

An acid–alkali reaction is a special case of an acid–base reaction, where the base used is also an alkali. When an acid reacts with an alkali salt (a metal hydroxide), the product is a metal salt and water. Acid–alkali reactions are also neutralization reactions.

In general, acid–alkali reactions can be simplified to

$$OH^-_{(aq)} + H^+_{(aq)} \rightarrow H_2O$$

by omitting spectator ions.

Acids are in general pure substances that contain hydrogen cations (H^+) or cause them to be produced in solutions. Hydrochloric acid (HCl) and sulfuric acid (H_2SO_4) are common examples. In water, these break apart into ions:

$$HCl \rightarrow H^+_{(aq)} + Cl^-_{(aq)}$$

$$H_2SO_4 \rightarrow H^+_{(aq)} + HSO^-_{4(aq)}$$

The alkali breaks apart in water, yielding dissolved hydroxide ions:

$$NaOH \rightarrow N^+_{(aq)} + OH^-_{(aq)}$$

The Arrhenius Theory

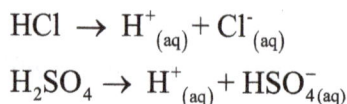

An acid is any hydrogen-containing compound that provides hydrogen ion in aqueous medium. Similarly, according to the theory, a base is any compound that gives hydroxyl ion in aqueous medium.

Example:

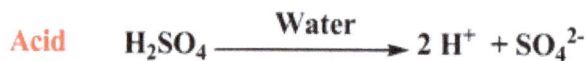

Acid $H_2SO_4 \xrightarrow{\text{Water}} 2\,H^+ + SO_4^{2-}$

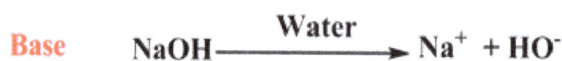

Base $NaOH \xrightarrow{\text{Water}} Na^+ + HO^-$

Merit:

A. With the help of this theory, the strength of acids and bases could be explained in terms of ionization equilibrium:

B. The existence of bare H^+ in aqueous medium was questioned. The size of proton is extremely small and therefore, it was proposed that it combined with water molecule and exists as H_3O^+ (hydronium) ion.

Franklin Theory (1905)

This theory concerns the acidic and basic properties of a substance in non-aqueous medium, mainly, in ammonium.

According to this theory,

An acid is a solute that gives the cationic character of the solvent, and a base is a solute that provides anionic character of the solvent.

Similar to aqueous medium,

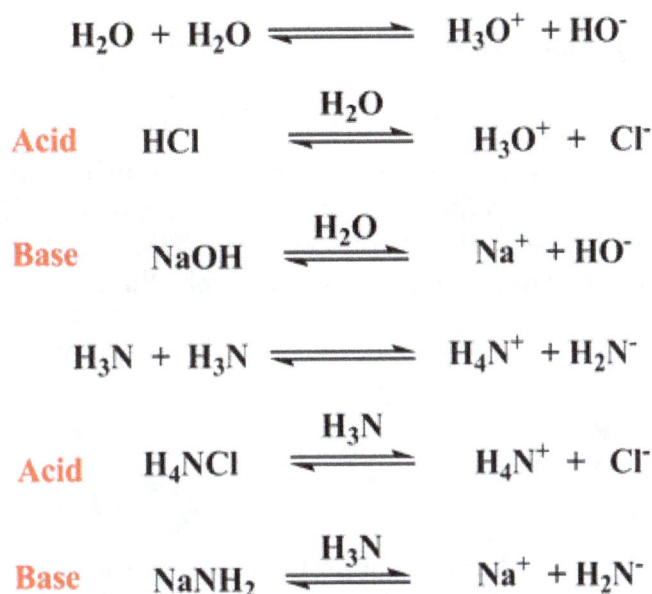

$$H_2O + H_2O \rightleftharpoons H_3O^+ + HO^-$$

Acid $HCl \xrightarrow{\text{H}_2\text{O}} \rightleftharpoons H_3O^+ + Cl^-$

Base $NaOH \xrightarrow{\text{H}_2\text{O}} \rightleftharpoons Na^+ + HO^-$

$$H_3N + H_3N \rightleftharpoons H_4N^+ + H_2N^-$$

Acid $H_4NCl \xrightarrow{\text{H}_3\text{N}} \rightleftharpoons H_4N^+ + Cl^-$

Base $NaNH_2 \xrightarrow{\text{H}_3\text{N}} \rightleftharpoons Na^+ + H_2N^-$

Merit and demerits:

Arrhenius theory was limited to aqueous medium, while, with the help of this theory acidic and basic properties of some substances in non-aqueous medium can be explained.

This definition also failed to define inherent acid-base character of a substance. Isotope labeling experiments showed that there was no dissociation and association of substance and solvent, hence, autoionization of solvents was questioned.

Brønsted–Lowry Acid–base Theory

The Brønsted–Lowry theory is an acid–base reaction theory which was proposed independently by Johannes Nicolaus Brønsted and Thomas Martin Lowry in 1923. The fundamental concept of this theory is that when an acid and a base react with each other, the acid forms its conjugate base, and the base forms its conjugate acid by exchange of a proton (the hydrogen cation, or H^+). This theory is a generalization of the Arrhenius theory.

Definitions of Acids and Bases

In the Arrhenius theory acids are defined as substances which dissociate in aqueous solution to give H^+ (hydrogen ions). Bases are defined as substances which dissociate in aqueous solution to give OH^- (hydroxide ions).

In 1923 physical chemists Johannes Nicolaus Brønsted in Denmark and Thomas Martin Lowry in England independently proposed the theory that carries their names. In the Brønsted–Lowry theory acids and bases are defined by the way they react with each other, which allows for greater generality. The definition is expressed in terms of an equilibrium expression

acid + base $\rightleftharpoons$ conjugate base + conjugate acid.

With an acid, HA, the equation can be written symbolically as:

$$HA + B \rightleftharpoons A^- + HB^+$$

The equilibrium sign, $\rightleftharpoons$, is used because the reaction can occur in both forward and backward directions. The acid, HA, can lose a proton to become its conjugate base, A^-. The base, B, can accept a proton to become its conjugate acid, HB^+. Most acid-base reactions are fast so that the components of the reaction are usually in dynamic equilibrium with each other.

Aqueous Solutions

Acetic acid, a weak acid, donates a proton (hydrogen ion, highlighted in green) to water in an equilibrium reaction to give the acetate ion and the hydronium ion. Red: oxygen, black: carbon, white: hydrogen.

Consider the following acid–base reaction:

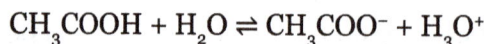

$$CH_3COOH + H_2O \rightleftharpoons CH_3COO^- + H_3O^+$$

Acetic acid, CH_3COOH, is an acid because it donates a proton to water (H_2O) and becomes its conjugate base, the acetate ion (CH_3COO^-). H_2O is a base because it accepts a proton from CH_3COOH and becomes its conjugate acid, the hydronium ion, (H_3O^+).

The reverse of an acid-base reaction is also an acid-base reaction, between the conjugate acid of the base in the first reaction and the conjugate base of the acid. In the above example, acetate is the base of the reverse reaction and hydronium ion is the acid.

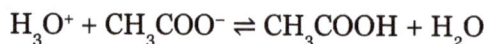

$$H_3O^+ + CH_3COO^- \rightleftharpoons CH_3COOH + H_2O$$

The power of the Brønsted–Lowry theory is that, in contrast to Arrhenius theory, it does not require an acid to dissociate.

Amphoteric Substances

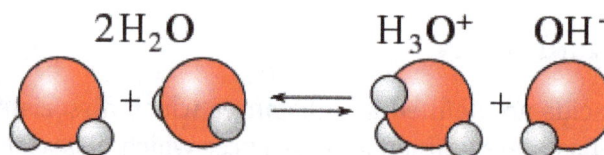

$$2H_2O \qquad H_3O^+ \quad OH^-$$

The amphoteric nature of water

The essence of Brønsted–Lowry theory is that an acid only exists as such in relation to a base, and *vice versa*. Water is amphoteric as it can act as an acid or as a base. In the image shown at the right one molecule of H_2O acts as a base and gains H^+ to become H_3O^+ while the other acts as an acid and loses H^+ to become OH^-.

Another example is furnished by substances like aluminium hydroxide, $Al(OH)_3$.

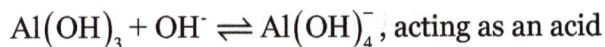

$$Al(OH)_3 + OH^- \rightleftharpoons Al(OH)_4^-, \text{ acting as an acid}$$

$$3H^+ + Al(OH)_3 \rightleftharpoons 3H_2O + Al^{3+}(aq), \text{ acting as a base}$$

Non-aqueous Solutions

The hydrogen ion, or hydronium ion, is a Brønsted–Lowry acid in aqueous solutions, and the hydroxide ion is a base, by virtue of the self-dissociation reaction

$$H_2O + H_2O \rightleftharpoons H_3O^+ + OH^-$$

An analogous reaction occurs in liquid ammonia

$$NH_3 + NH_3 \rightleftharpoons NH_4^+ + NH_2^-$$

Thus, the ammonium ion, NH_4^+, plays the same role in liquid ammonia as does the hydronium ion in water and the amide ion, NH_2^-, is analogous to the hydroxide ion. Ammonium salts behave as acids, and amides behave as bases.

Some non-aqueous solvents can behave as bases, that is, proton acceptors, in relation to Brønsted–Lowry acids.

$$HA + S \rightleftharpoons A^- + SH^+$$

where S stands for a solvent molecule. The most important such solvents are dimethylsulfoxide, DMSO, and acetonitrile, CH_3CN, as these solvents has been widely used to measure the acid dissociation constants of organic molecules. Because DMSO is a stronger proton acceptor than H_2O the acid becomes a stronger acid in this solvent than in water. Indeed, many molecules behave as acids in non-aqueous solution that do not do so in aqueous solution. An extreme case occurs with carbon acids, where a proton is extracted from a C–H bond.

Some non-aqueous solvents can behave as acids. An acidic solvent will increase basicity of substances dissolved in it. For example, the compound CH_3COOH is known as acetic acid because of its acidic behaviour in water. However it behaves as a base in liquid hydrogen chloride, a much more acidic solvent.

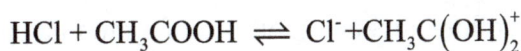

$$HCl + CH_3COOH \rightleftharpoons Cl^- + CH_3C(OH)_2^+$$

Comparison with Lewis Acid–base Theory

In the same year that Brønsted and Lowry published their theory, G. N. Lewis proposed an alternative theory of acid–base reactions. The Lewis theory is based on electronic structure. A Lewis base is defined as a compound that can donate an electron pair to a Lewis acid, a compound that can accept an electron pair. Lewis's proposal gives an explanation to the Brønsted–Lowry classification in terms of electronic structure.

$$HA + B: \rightleftharpoons A:^- + BH^+$$

In this representation both the base, B, and the conjugate base, A⁻, are shown carrying a lone pair of electrons and the proton, which is a Lewis acid, is transferred between them.

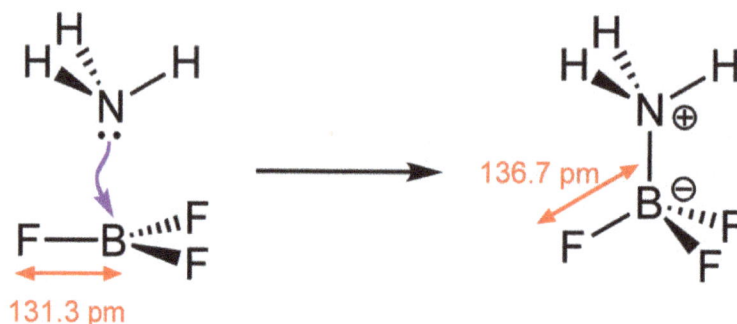

Adduct of ammonia and boron trifluoride

Lewis later wrote in "To restrict the group of acids to those substances that contain hydrogen interferes as seriously with the systematic understanding of chemistry as would the restriction of the term oxidizing agent to substances containing oxygen." In Lewis theory an acid, A, and a base, B:, form an adduct, AB, in which the electron pair is used to form a dative covalent bond between A and B. This is illustrated with the formation of the adduct $H_3N–BF_3$ from ammonia and boron trifluoride, a reaction that cannot occur in aqueous solution because boron trifluoride reacts violently with water in a hydrolysis reaction.

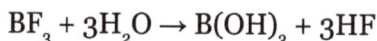

$$BF_3 + 3H_2O \rightarrow B(OH)_3 + 3HF$$

$$HF \rightleftharpoons H^+ + F^-$$

These reactions illustrate that BF_3 is an acid in both Lewis and Brønsted–Lowry classifications and emphasizes the consistency between both theories.

Boric acid is recognized as a Lewis acid by virtue of the reaction

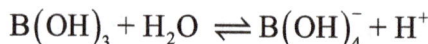

$$B(OH)_3 + H_2O \rightleftharpoons B(OH)_4^- + H^+$$

In this case the acid does not dissociate, it is the base, H_2O that dissociates. A solution of $B(OH)_3$ is acidic because hydrogen ions are liberated in this reaction.

There is strong evidence that dilute aqueous solutions of ammonia contain negligible amounts of the ammonium ion

$$H_2O + NH_3 \rightarrow OH^- + NH_4^+$$

and that, when dissolved in water, ammonia functions as a Lewis base.

Comparison with the Lux-Flood Theory

The reactions between certain oxides in non-aqueous media cannot be explained on the basis of Brønsted–Lowry theory. For example, the reaction

$$2MgO + SiO_2 \rightarrow Mg_2SiO_4$$

does not fall within the scope of the Brønsted–Lowry definition of acids and bases. On the other hand, MgO is basic and SiO_2 is acidic in the Brønsted–Lowry sense, referring to mixtures in water.

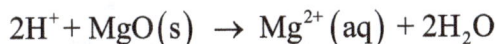

$$2H^+ + MgO(s) \rightarrow Mg^{2+}(aq) + 2H_2O$$

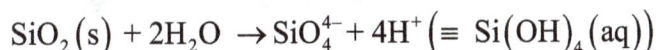

$$SiO_2(s) + 2H_2O \rightarrow SiO_4^{4-} + 4H^+ \left(\equiv Si(OH)_4(aq) \right)$$

Lux-Flood theory also classifies magnesium oxide as a base in non-aqueous circumstances. This classification is important in geochemistry. Minerals such as olivine, $(Mg,Fe)SiO_4$ are classed as ultramafic; olivine is a compound of a very basic oxide, MgO, with an acidic oxide, silica, SiO_2.

Protonic Theory

Any species that tends to give up a proton is an acid, and any species that tends to accept a proton is a base.

Examples:

$$HCl + H_2O \rightleftharpoons H_3O^+ + Cl^-$$

Acid   Base

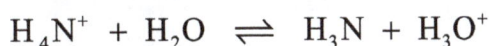

$$H_4N^+ + H_2O \rightleftharpoons H_3N + H_3O^+$$

Acid   Base

Conjugate acid-base pair:

Two chemical species, which are inter-convertible to each other by means of a proton, are known as conjugate acid-base pairs.

Example:

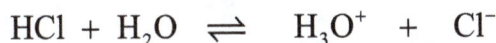

$$HCl + H_2O \rightleftharpoons H_3O^+ + Cl^-$$

Acid   Base   Conjugate acid Conjugate base

$$H_4N^+ + H_2O \rightleftharpoons H_3N + H_3O^+$$

Acid   Base Conjugate base Conjugate acid

A species, which has high tendency to give up proton(s) in a medium, will behave as a strong acid. The corresponding conjugate base, then, must be weak in nature, i.e. its tendency for accepting proton(s) will be very low. The strength of an acid-species is also dependent to the stability of the corresponding conjugate base. The higher stability of the conjugate base favors the deprotonation process by the easy release of proton.

Example:

$$HCl + H_2O \rightleftharpoons H_3O^+ + Cl^-$$

$$\text{Acid} \quad \text{Base} \quad \text{Conjugate acid} \quad \text{Conjugate base}$$

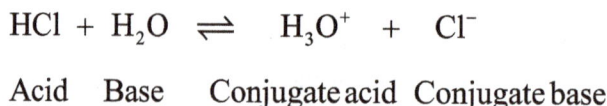

Similarly, a conjugate acid of a strong base should be weak in nature.

Example:

$$H_4N^+ + H_2O \rightleftharpoons H_3N + H_3O^+$$

$$\text{Acid} \quad \text{Base} \quad \text{Conjugate base} \quad \text{Conjugate acid}$$

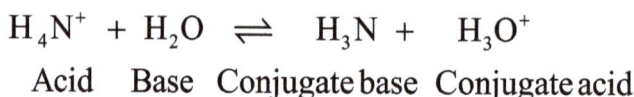

Merits and demerits:

Merits:

Acidic-basic properties of substances can be explained in any protonic solvent such as liquid ammonia, sulphuric acid, etc.

With the help of this theory the strength of acids and bases can be calculated. Further more, the reason that tunes the strength of acidic and basic properties (conjugate acid-base theory) can be understood.

Demerit:

In this theory, the acid-base properties are determined solely by proton exchange parameter. Therefore, some other substances that do not contain any proton but have inherent acidic or basic properties, e.g. BF_3, I_2, $AlCl_3$, etc cannot be explained with the help of the theory.

Lewis Acids and Bases

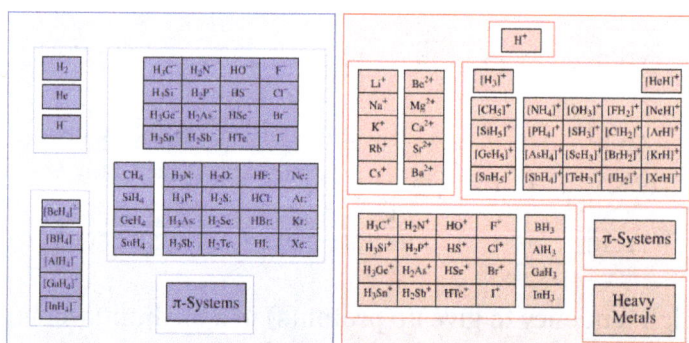

Diagram of some Lewis bases and acids

A Lewis acid is a chemical species that reacts with a Lewis base to form a Lewis adduct. A Lewis base, then, is any species that donates an unshared electron pair to a Lewis acid to form a Lewis adduct. For example, OH^- and NH_3 are Lewis bases, because they can donate a lone pair of electrons. In the adduct, the Lewis acid and base share an electron pair furnished by the Lewis base. Usually the terms *Lewis acid* and *Lewis base* are defined within the context of a specific chemical reaction. For example, in the reaction of Me_3B and NH_3 to give Me_3BNH_3, Me_3B acts as a Lewis acid, and NH_3 acts as a Lewis base. Me_3BNH_3 is the Lewis adduct. The terminology refers to the contributions of Gilbert N. Lewis.

Depicting adducts

In many cases, the interaction between the Lewis base and Lewis acid in a complex is indicated by an arrow indicating the Lewis base donating electrons toward the Lewis acid using the notation of a dative bond—for example, $Me_3B{\leftarrow}NH_3$. Some sources indicate the Lewis base with a pair of dots (the explicit electrons being donated), which allows consistent representation of the transition from the base itself to the complex with the acid:

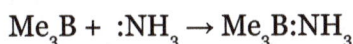

$$Me_3B + {:}NH_3 \rightarrow Me_3B{:}NH_3$$

A center dot may also be used to represent a Lewis adduct, such as $Me_3B{\cdot}NH_3$. Another example is boron trifluoride diethyl etherate, $BF_3{\cdot}Et_2O$. The center dot is also used to represent hydrate coordination in various crystals, as in $MgSO_4{\cdot}7H_2O$ for hydrated magnesium sulfate. In general, however, the donor–acceptor bond is viewed as simply somewhere along a continuum between idealized covalent bonding and ionic bonding.

Examples

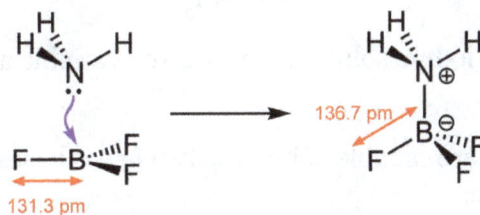

Major structural changes accompany binding of the Lewis base to the coordinatively unsaturated, planar Lewis acid BF_3.

Classically, the term "Lewis acid" is restricted to trigonal planar species with an empty p orbital, such as BR_3 where R can be an organic substituent or a halide. For the purposes of discussion, even complex compounds such as $Et_3Al_2Cl_3$ and $AlCl_3$ are treated as trigonal planar Lewis acids. Metal ions such as Na^+, Mg^{2+}, and Ce^{3+}, which are invariably complexed with additional ligands, are often sources of coordinatively unsaturated derivatives that form Lewis adducts upon reaction with a Lewis base. Other reactions might simply be referred to as "acid-catalyzed" reactions. Some compounds, such as H_2O, are both Lewis acids and Lewis bases, because they can either accept a pair of electrons or donate a pair of electrons, depending upon the reaction.

Lewis acids are diverse. Simplest are those that react directly with the Lewis base. But more common are those that undergo a reaction prior to forming the adduct.

- Examples of Lewis acids based on the general definition of electron pair acceptor include:

  o the proton (H^+) and acidic compounds onium ions, such as NH_4^+ and H_3O^+

  o metal cations, such as Li^+ and Mg^{2+}, often as their aquo or ether complexes,

  o trigonal planar species, such as BF_3 and carbocations H_3C^+

  o pentahalides of phosphorus, arsenic, and antimony

  o electron poor π-systems, such as enones and tetracyanoethylenes.

Again, the description of a Lewis acid is often used loosely. For example, in solution, bare protons do not exist.

Simple Lewis Acids

The most studied examples of such Lewis acids are the boron trihalides and organoboranes, but other compounds exhibit this behavior:

$$BF_3 + F^- \rightarrow BF_4^-$$

In this adduct, all four fluoride centres (or more accurately, ligands) are equivalent.

$$BF_3 + OMe_2 \rightarrow BF_3OMe_2$$

Both BF_4^- and BF_3OMe_2 are Lewis base adducts of boron trifluoride.

In many cases, the adducts violate the octet rule, such as the triiodide anion:

$$I_2 + I^- \rightarrow I_3^-$$

The variability of the colors of iodine solutions reflects the variable abilities of the solvent to form adducts with the Lewis acid I_2.

In some cases, the Lewis acids are capable of binding two Lewis bases, a famous example being the formation of hexafluorosilicate:

$$SiF_4 + 2\ F^- \rightarrow SiF_6^{2-}$$

Complex Lewis Acids

Most compounds considered to be Lewis acids require an activation step prior to formation of the adduct with the Lewis base. Well known cases are the aluminium trihalides, which are widely viewed as Lewis acids. Aluminium trihalides, unlike the boron trihalides, do not exist in the form AlX_3, but as aggregates and polymers that must be degraded by the Lewis base. A simpler case is the formation of adducts of borane. Monomeric BH_3 does not exist appreciably, so the adducts of borane are generated by degradation of diborane:

$$B_2H_6 + 2\ H^- \rightarrow 2\ BH_4^-$$

In this case, an intermediate $B_2H_7^-$ can be isolated.

Many metal complexes serve as Lewis acids, but usually only after dissociating a more weakly bound Lewis base, often water.

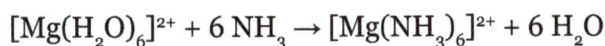

$$[Mg(H_2O)_6]^{2+} + 6\,NH_3 \rightarrow [Mg(NH_3)_6]^{2+} + 6\,H_2O$$

H^+ as Lewis Acid

The proton (H^+) is one of the strongest but is also one of the most complicated Lewis acids. It is convention to ignore the fact that a proton is heavily solvated (bound to solvent). With this simplification in mind, acid-base reactions can be viewed as the formation of adducts:

- $H^+ + NH_3 \rightarrow NH_4^+$

- $H^+ + OH^- \rightarrow H_2O$

Applications of Lewis Acids

A typical example of a Lewis acid in action is in the Friedel–Crafts alkylation reaction. The key step is the acceptance by $AlCl_3$ of a chloride ion lone-pair, forming $AlCl_4^-$ and creating the strongly acidic, that is, electrophilic, carbonium ion.

$$RCl + AlCl_3 \rightarrow R^+ + AlCl_4^-$$

Lewis Bases

A Lewis base is an atomic or molecular species where the highest occupied molecular orbital (HOMO) is highly localized. Typical Lewis bases are conventional amines such as ammonia and alkyl amines. Other common Lewis bases include pyridine and its derivatives. Some of the main classes of Lewis bases are

- amines of the formula $NH_{3-x}R_x$ where R = alkyl or aryl. Related to these are pyridine and its derivatives.

- phosphines of the formula $PR_{3-x}A_x$, where R = alkyl, A = aryl.

- compounds of O, S, Se and Te in oxidation state 2, including water, ethers, ketones

The most common Lewis bases are anions. The strength of Lewis basicity correlates with the pK_a of the parent acid: acids with high pK_a's give good Lewis bases. As usual, a weaker acid has a stronger conjugate base.

- Examples of Lewis bases based on the general definition of electron pair donor include:

  o simple anions, such as H^- and F^-.

  o other lone-pair-containing species, such as H_2O, NH_3, HO^-, and CH_3^-

  o complex anions, such as sulfate

  o electron rich π-system Lewis bases, such as ethyne, ethene, and benzene

The strength of Lewis bases have been evaluated for various Lewis acids, such as I_2, $SbCl_5$, and BF_3.

Heats of binding of various bases to BF$_3$		
Lewis base	**donor atom**	**Enthalpy of Complexation (kJ/mol)**
Et$_3$N	N	135
quinuclidine	N	150
pyridine	N	128
Acetonitrile	N	60
Et$_2$O	O	78.8
THF	O	90.4
acetone	O	76.0
EtOAc	O	75.5
DMA	O	112
DMSO	O	105
Tetrahydrothiophene	S	51.6
Trimethylphosphine	P	97.3

Applications of Lewis Bases

Nearly all electron pair donors that form compounds by binding transition elements can be viewed as a collections of the Lewis bases – or ligands. Thus a large application of Lewis bases is to modify the activity and selectivity of metal catalysts. Chiral Lewis bases thus confer chirality on a catalyst, enabling asymmetric catalysis, which is useful for the production of pharmaceuticals.

Many Lewis bases are "multidentate," that is they can form several bonds to the Lewis acid. These multidentate Lewis bases are called chelating agents.

Hard and Soft Classification

Lewis acids and bases are commonly classified according to their hardness or softness. In this context hard implies small and nonpolarizable and soft indicates larger atoms that are more polarizable.

- typical hard acids: H$^+$, alkali/alkaline earth metal cations, boranes, Zn^{2+}

- typical soft acids: Ag$^+$, Mo(0), Ni(0), Pt^{2+}

- typical hard bases: ammonia and amines, water, carboxylates, fluoride and chloride

- typical soft bases: organophosphines, thioethers, carbon monoxide, iodide

For example, an amine will displace phosphine from the adduct with the acid BF$_3$. In the same way, bases could be classified. For example, bases donating a lone pair from an oxygen atom are harder than bases donating through a nitrogen atom. Although the classification was never quantified it proved to be very useful in predicting the strength of adduct formation, using the key concepts that hard acid — hard base and soft acid — soft base interactions are stronger than hard acid — soft base or soft acid — hard base interactions. Later investigation of the thermodynamics of the interaction suggested that hard—hard interactions are enthalpy favored, whereas soft—soft are entropy favored.

There is no Single Order of Lewis Base Strengths

Cramer and Bopp have shown graphically using the E and C parameters of the ECW Model that there is no one single order of Lewis base strengths (or acid strengths). Single property or variable scales are limited to a small range of acids or bases.

History

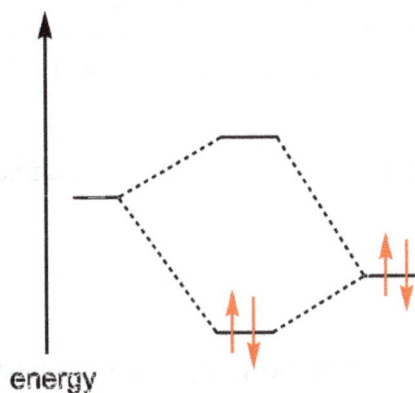

MO diagram depicting the formation of a dative covalent bond between two atoms.

The concept originated with Gilbert N. Lewis who studied chemical bonding. In 1923, Lewis wrote *An acid substance is one which can employ an electron lone pair from another molecule in completing the stable group of one of its own atoms.* The Brønsted–Lowry acid–base theory was published in the same year. The two theories are distinct but complementary. A Lewis base is also a Brønsted–Lowry base, but a Lewis acid doesn't need to be a Brønsted–Lowry acid. The classification into hard and soft acids and bases (HSAB theory) followed in 1963. The strength of Lewis acid-base interactions, as measured by the standard enthalpy of formation of an adduct can be predicted by the Drago–Wayland two-parameter equation.

Reformulation of Lewis theory

Lewis had suggested in 1916 that two atoms are held together in a chemical bond by sharing a pair of electrons. When each atom contributed one electron to the bond it was called a covalent bond. When both electrons come from one of the atoms it was called a dative covalent bond or coordinate bond. The distinction is not very clear-cut. For example, in the formation of an ammonium ion from ammonia and hydrogen the ammonia molecule donates a pair of electrons to the proton; the identity of the electrons is lost in the ammonium ion that is formed. Nevertheless, Lewis suggested that an electron-pair donor be classified as a base and an electron-pair acceptor be classified as acid.

A more modern definition of a Lewis acid is an atomic or molecular species with a localized empty atomic or molecular orbital of low energy. This lowest energy molecular orbital (LUMO) can accommodate a pair of electrons.

Comparison with Brønsted–Lowry Theory

A Lewis base is often a Brønsted–Lowry base as it can donate a pair of electrons to H^+; the proton is a Lewis acid as it can accept a pair of electrons. The conjugate base of a Brønsted–Lowry acid is also a Lewis base as loss of H^+ from the acid leaves those electrons which were used for the A—H bond as a lone pair on the conjugate base. However, a Lewis base can be very difficult to protonate, yet still react with a Lewis acid. For example, carbon monoxide is a very weak Brønsted–Lowry base but it forms a strong adduct with BF_3.

In another comparison of Lewis and Brønsted–Lowry acidity by Brown and Kanner, 2,6-di-*t*-butylpyridine reacts to form the hydrochloride salt with HCl but does not react with BF_3. This example demonstrates that steric factors, in addition to electron configuration factors, play a role in determining the strength of the interaction between the bulky di-*t*-butylpyridine and tiny proton.

A Brønsted–Lowry acid is a proton donor, not an electron-pair acceptor.

Merits and Demerits of Lewis Acid-base

Merits:

With the help of this concept acid-base property of both Brønsted-type and nonprotonic substances can be explained. Besides that, with the help of this theory acid-base character and oxidation-reduction power of a substance can be postulated. For example, Lewis base donates electron pair, which is equivalent to reduction. Therefore, with increase of Lewis base character reducing power should increase. Lewis acid accepts lone pair of electrons, therefore, it is an oxidizing agent.

Demerits:

i.   The conventional protonic acids, like H_2SO_4, HCl, HNO_3, etc are not covered directly by this theory, because no covalent bond is formed by between accepting a pair of electrons.

ii.   The strength of acid-base cannot be generalized by the theory. For example, fluoride complex of beryllium(II) is more stable than that of fluoride complex of copper(II). Therefore, beryllium(II) should be more Lewis acidic that copper(II). On the other hand, the ammonium complex of copper(II) is more stable that beryllium(II). This indicates copper(II) is more acidic that beryllium(II).

iii. The reaction catalyzed by Lewis acid is not catalyzed by protonic acid.

Lux-Flood & The Usanovich

Lux-Flood Definition (1939):

According to this theory, acid and base is define as, acid is an oxide acceptor and base is an oxide donor.

Example:

Here, CaO donates oxygen, therefore, it is a basic oxide. SiO_2 is an oxide acceptor, hence, it is a acidic oxide.

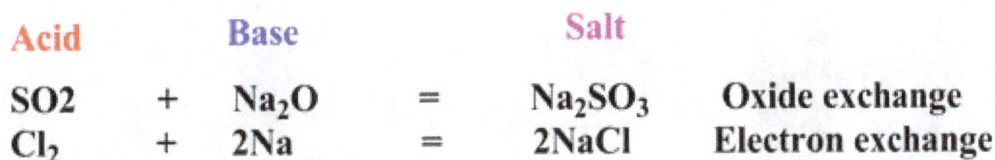

Merits and demerits:

Merit:

This approach emphasizes the acid and basic-anhydride aspects. Acidic oxides are acid anhydride and in the aqueous medium generate protonic acids, while, basic oxides are basic anhydride and generate hydroxyl ion in the aqueous medium.

Demerit:

The usefulness of this concept is limited to inorganic metal oxide and their reactions in molten state.

The Usanovich Definition (1939):

The Usanovich definition includes all reactions of Lewis acids and bases and extended the latter concept by removing the restriction that the donation or acceptance of electrons be as shared pairs. According to him,

an acid is any chemical species which (i) reacts with a base, or (ii) accept anions or electrons, or (iii) furnish cations; and a base is any chemical species which (i) reacts with acids, or (ii) gives up anions or electrons, or (iii) combines with cations.

Examples:

Acid		Base		Salt	
$SO2$	$+$	Na_2O	$=$	Na_2SO_3	Oxide exchange
Cl_2	$+$	$2Na$	$=$	$2NaCl$	Electron exchange

Merits and demerits:

Merits:

Lewis concept and protonic concepts have been covered under the Usanovis definition for acid-base. Besides that it states that the donation or acceptance of electrons need not take place as shared pairs. Accordingly, oxidation-reduction reactions may be classified as acid-base reaction.

Demerits:

This concept is extremely general and therefore, almost all reactions can be considered as acid-base reaction. Because of this, the convenience of treating a particular type of compounds as acids or bases has been lost.

HSAB Theory

HSAB concept is an initialism for "hard and soft (Lewis) acids and bases". Also known as the Pearson acid base concept, HSAB is widely used in chemistry for explaining stability of compounds, reaction mechanisms and pathways. It assigns the terms 'hard' or 'soft', and 'acid' or 'base' to chemical species. 'Hard' applies to species which are small, have high charge states (the charge criterion applies mainly to acids, to a lesser extent to bases), and are weakly polarizable. 'Soft' applies to species which are big, have low charge states and are strongly polarizable. The concept is a way of applying the notion of orbital overlap to specific chemical cases.

The theory is used in contexts where a qualitative, rather than quantitative, description would help in understanding the predominant factors which drive chemical properties and reactions. This is especially so in transition metal chemistry, where numerous experiments have been done to determine the relative ordering of ligands and transition metal ions in terms of their hardness and softness.

HSAB theory is also useful in predicting the products of metathesis reactions. In 2005 it was shown that even the sensitivity and performance of explosive materials can be explained on basis of HSAB theory.

Ralph Pearson introduced the HSAB principle in the early 1960s as an attempt to unify inorganic and organic reaction chemistry.

Theory

Hard-soft Trends for Acids and Bases

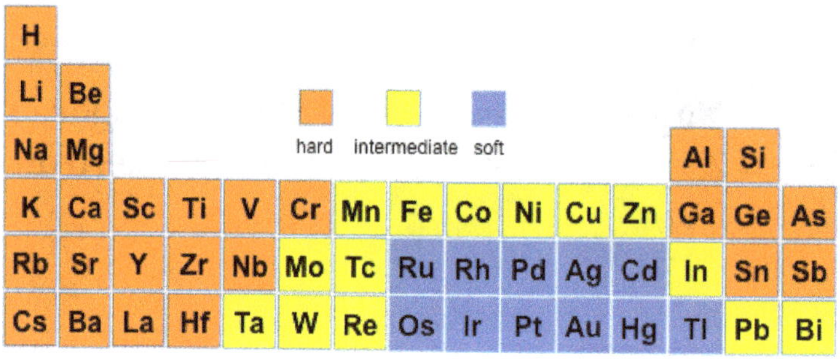

Acids

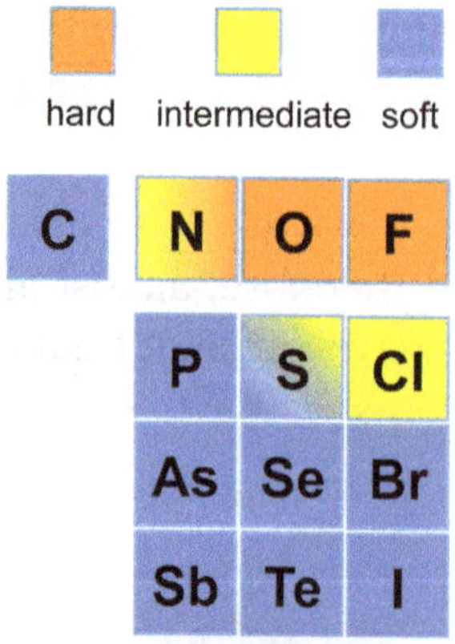

Bases

The gist of this theory is that *soft* acids react faster and form stronger bonds with *soft* bases, whereas *hard* acids react faster and form stronger bonds with *hard* bases, all other factors being equal.

The classification in the original work was mostly based on equilibrium constants for reaction of two Lewis bases competing for a Lewis acid.

Comparing tendencies of hard acids and bases vs. soft acids and bases		
Property	Hard acids and bases	Soft acids and bases
atomic/ionic radius	small	large
oxidation state	high	low or zero
polarizability	low	high
electronegativity (bases)	high	low
HOMO energy of bases	low	higher
LUMO energy of acids	high	lower (but > soft-base HOMO)
affinity	ionic bonding	covalent bonding

Examples of hard and soft acids and bases							
Acids				Bases			
hard		soft		hard		soft	
Hydronium	H_3O^+	Mercury	CH_3Hg^+, Hg^{2+}, Hg_2^{2+}	Hydroxide	OH^-	Hydride	H^-
Alkali metals	Li^+, Na^+, K^+	Platinum	Pt^{2+}	Alkoxide	RO^-	Thiolate	RS^-
Titanium	Ti^{4+}	Palladium	Pd^{2+}	Halogens	F^-, Cl^-	Halogens	I^-
Chromium	Cr^{3+}, Cr^{6+}	Silver	Ag^+	Ammonia	NH_3	Phosphine	PR_3
Boron trifluoride	BF_3	Borane	BH_3	Carboxylate	CH_3COO^-	Thiocyanate	SCN^-
Carbocation	R_3C^+	P-chloranil		Carbonate	CO_3^{2-}	Carbon monoxide	CO
Lanthanides	Ln^{3+}	Bulk metals	M^0	Hydrazine	N_2H_4	Benzene	C_6H_6
Thorium	Th^{4+}	Gold	Au^+				

Borderline cases are also identified: borderline acids are trimethylborane, sulfur dioxide and ferrous Fe^{2+}, cobalt Co^{2+} caesium Cs^+ and lead Pb^{2+} cations. Borderline bases are: aniline, pyridine, nitrogen N_2 and the azide, chloride, bromide, nitrate and sulfate anions.

Generally speaking, acids and bases interact and the most stable interactions are hard-hard (ionogenic character) and soft-soft (covalent character).

An attempt to quantify the 'softness' of a base consists in determining the equilibrium constant for the following equilibrium:

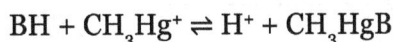

$$BH + CH_3Hg^+ \rightleftharpoons H^+ + CH_3HgB$$

Where CH_3Hg^+ (methylmercury ion) is a very soft acid and H^+ (proton) is a hard acid, which compete for B (the base to be classified).

Some examples illustrating the effectiveness of the theory:

- Bulk metals are soft acids and are poisoned by soft bases such as phosphines and sulfides.

- Hard solvents such as hydrogen fluoride, water and the protic solvents tend to solvate strong solute bases such as the fluorine anion and the oxygen anions. On the other hand, dipolar aprotic solvents such as dimethyl sulfoxide and acetone are soft solvents with a preference for solvating large anions and soft bases.

- In coordination chemistry soft-soft and hard-hard interactions exist between ligands and metal centers.

Chemical Hardness

Chemical hardness in electron volt					
Acids			Bases		
Hydrogen	H^+	∞	Fluoride	F^-	7
Aluminium	Al^{3+}	45.8	Ammonia	NH_3	6.8
Lithium	Li^+	35.1	hydride	H^-	6.8
Scandium	Sc^{3+}	24.6	carbon monoxide	CO	6.0
Sodium	Na^+	21.1	hydroxyl	OH^-	5.6
Lanthanum	La^{3+}	15.4	cyanide	CN^-	5.3
Zinc	Zn^{2+}	10.8	phosphane	PH_3	5.0
Carbon dioxide	CO_2	10.8	nitrite	NO_2^-	4.5
Sulfur dioxide	SO_2	5.6	Hydrosulfide	SH^-	4.1
Iodine	I_2	3.4	Methane	CH_3^-	4.0
Chemical hardness data					

In 1983 Pearson together with Robert Parr extended the qualitative HSAB theory with a quantitative definition of the chemical hardness (η) as being proportional to the second derivative of the total energy of a chemical system with respect to changes in the number of electrons at a fixed nuclear environment:

$$\eta = \frac{1}{2}\left(\frac{\partial^2 E}{\partial N^2}\right)_Z.$$

The factor of one-half is arbitrary and often dropped as Pearson has noted.

An operational definition for the chemical hardness is obtained by applying a three-point finite difference approximation to the second derivative:

$$\approx \frac{E(N+1)-2E(N)+E(N-1)}{}$$

$$\frac{(E(N-1)-E(N))-(E(N)-E(N+1))}{}$$

$$= -(I-A)$$

where I is the ionization potential and A the electron affinity. This expression implies that the chemical hardness is proportional to the band gap of a chemical system, when a gap exists.

The first derivative of the energy with respect to the number of electrons is equal to the chemical potential, μ, of the system,

$$\mu = \left(\frac{\partial E}{\partial N}\right)_z,$$

from which an operational definition for the chemical potential is obtained from a finite difference approximation to the first order derivative as

$$\mu \approx \frac{E(N+1) - E(N-1)}{2}$$
$$= \frac{-(E(N-1) - E(N)) - (E(N) - E(N+1))}{2}$$
$$= -\frac{1}{2}(I + A)$$

which is equal to the negative of the electronegativity (χ) definition on the Mulliken scale: $\mu = -\chi$.

The hardness and Mulliken electronegativity are related as

$$2\eta = \left(\frac{\partial \mu}{\partial N}\right)_z \approx -\left(\frac{\partial \chi}{\partial N}\right)_z,$$

and in this sense hardness is a measure for resistance to deformation or change. Likewise a value of zero denotes maximum softness, where softness is defined as the reciprocal of hardness.

In a compilation of hardness values only that of the hydride anion deviates. Another discrepancy noted in the original 1983 article are the apparent higher hardness of Tl^{3+} compared to Tl^{+}.

Modifications

If the interaction between acid and base in solution results in an equilibrium mixture the strength of the interaction can be quantified in terms of an equilibrium constant. An alternative quantitative measure is the standard heat (enthalpy) of formation of the adduct in a non-coordinating solvent. A two-parameter equation that predicts the formation of a large number of adducts is the ECW Model.

$$-\Delta H^{\ominus}(A - B) = E_A E_B + C_A C_B$$

Value of the E and C parameters have been tabulated. Related E and C analyses quantitatively predict of formation constants for complexes of many metal ions plus the proton with a wide range of unidentate Lewis acids in aqueous solution, and also offered insights into factors governing HSAB behavior in solution.

Another quantitative system has been proposed, in which Lewis acid strength is based on gas-phase affinity for fluoride.

Kornblum's Rule

An application of HSAB theory is the so-called Kornblum's rule which states that in reactions with ambient nucleophiles (nucleophiles that can attack from two or more places), the more electronegative atom reacts when the reaction mechanism is S_N1 and the less electronegative one in a S_N2 reaction. This rule (established in 1954) predates HSAB theory but in HSAB terms its explanation is that in a S_N1 reaction the carbocation (a hard acid) reacts with a hard base (high electronegativity) and that in a S_N2 reaction tetravalent carbon (a soft acid) reacts with soft bases.

According to findings, electrophilic alkylations at free CN^- occur preferentially at carbon, regardless of whether the S_N1 or S_N2 mechanism is involved and whether hard or soft electrophiles are employed. Preferred N attack, as postulated for hard electrophiles by the HSAB principle, could not be observed with any alkylating agent. Isocyano compounds are only formed with highly reactive electrophiles that react without an activation barrier because the diffusion limit is approached. It is claimed that the knowledge of absolute rate constants and not of the hardness of the reaction partners is needed to predict the outcome of alkylations of the cyanide ion.

Criticism

In 2011, Herbert Mayr et al. from Ludwig Maximilian University of Munich (LMU) published a critical review in Angewandte Chemie. Consecutive analysis of various types of ambient organic system reveals that an older approach based on thermodynamic/kinetic control describes reactivity of organic compounds perfectly, while the HSAB principle actually fails and should be abandoned in the rationalization of ambient reactivity of organic compounds.

Strength of Lewis Acids and Bases

Strength of Lewis acids and bases [Hard (H) and Soft (S) Acid (A) Base (B) Concept = HSAB Concept), 1963]:

Lewis defines acids are electrons acceptor and bases are electron donor. Therefore, the strength of acids and bases is determined by the nature of electron transfer in a particular reaction. Hence, the strength is dependent on a particular reaction. Accordingly, assignment of any single consistent criterion for acid-base strength becomes very difficult in the Lewis definition. However, a qualitative correlation between the various Lewis acids and Bases has been obtained by classifying the acids and bases in to two different groups, known as hard and soft.

In 1963, R. G. Pearson proposed that hard acids prefer to combine with hard bases and soft acids prefer to combine with soft bases. This is known as HSAB concept or theory.

The features distinguish Hard and Soft acids:

Hard acid	Soft acid
(I) Small in size.	(I) Large in Size
(II) High positive oxidation state.	(II) Zero or low positive oxidation state.
(III) Absence of any outer electrons that are easily exited to higher states.	(III) Sevral easily excitable valence electrons.

The features distinguish Hard and Soft bases:

Hard bases	Soft bases
(I) High electronegativity.	(I) Low electronegativity.
(II) Low polarisability.	(II) High polarisability.
(III) Presence of filled orbitals; empty orbitals may exist at high energy level.	(III) Presence of partially filled orbitals; empty orbitals are low-lying.

Table showing strength of HSAB, and border line case:

Bases(Nucleophiles)	Acids(Electrophiles)
Hard	Hard
H_2O, OH^-, F^-	H^+, Li^+, Na^+, K^+
$CH_2CO_2^-, PO_4^{2-}, SO_4^{2-}$	$Be^{2+}, Mg^{2+}, Ca^{2+}$
$Cl^-, CO_3^{2-}, ClO_4, NO_3^-$	Al^{3+}, Ga^{3+}
ROH, RO^-, R_2O	$Cr^{3+}, Co^{3+}, Fe^{3+},$
NH_3, RNH_2, N_2H_4	CH_3Sn^{3+}
	Si^{4+}, Ti^{4+}
	Ce^{3+}, Sn^{4+}
	$(CH_3)_2, Sn^{2+}$
	$BeMe_2, BF, B(OR)_3$
	$Al(CH_3)_3, AlCl_3, AlH_3$
	$RPO_2^+, ROPO_2^+$
	$RSO_2^+, ROSO_2^+, SO_3$
	$I^{7+}, I^{3+}, Cl^{7+}, Cr^{6+}$
	RCO^+, CO_3, NC^+
	HX(hydrogen bonding molocules)
Soft	*Soft*
R_2S, RSH, RS^-	$Cu^+, Ag^+, Au^+, Tl^+, Hg^+$
$I^-, SCN^-, S_2O_3^{2-}$	$Pd^{2+}, Cd^{2+}, Pt^{2+}, Hg^+, CH_3Hg^+$
$R_3P, R_3As, (RO)_3P$	$Co(CN)_5^{2-}$
CN^-, RNC, CO	$Tl^{3+}, Tl(CH_3)_3, BH_3$
C_2H_4, C_6H_6	Rs^+, RSe^+, RTe^+
H^-, R^-	I^+, Br^+, HO^+, RO^+
	$I_2, Br_2, ICN, etc.$

	trinitrobenzene,etc.
	chloranil,quinones,etc.
	tetracyanoethylene,etc.
	O,Cl,Br,I,N,RO,RO_2
	$M°$ (metal atoms)
	bulk metal
	CH_2,carbenes
Borderline	*Borderline*
$C_6H_5NH_2,C_5H_5N,N_3^-,Br^-,NO_2^-,$ SO_3^{2-},N_2	$Fe^{2+},Co^{2+},Ni^{2+},Cu^{2+},Zn^{2+},Pb^{2+},$ $Sn^{2+},B(CH_3)_3,SO_2,NO^+,R_3C^+,$ $C_6H_5^+$

General trends in acid strength:

Protonic acids can be classified in two groups, (A) hydro-acids, and (B) oxo-acids.

(A) Hydro-acids: In hydro-acids proton(s) is(are) directly attached to some second element. In the periodic table the acidity of hydro-acids increases from left to right along a period and top to bottom along a group.

Example: $HF > H_2O > NH_3$

Here, on going from left to right in a period, electronegativity of the elements attached to proton increases, therefore, the proton-element bond become more polar. As the polarity of the bond increases, release of proton becomes easier. Hence, acidity increases.

Example: $HI > HBr > HCl > HF$

In this case, the size of the elements increases and electronegativity decreases on moving from top to bottom in a group. Here, size is the dominating factor over electronegativity. As size increases, the extent of overlap between small proton and big elements becomes low and proton-element bond becomes weak. Therefore, the release of proton becomes easier as the size of element increases, consequently, acidity increases.

(B) Strength of oxy-acids: In case of oxo acid of type X - O - H, the acidity of the acid decreases with lower the position of X in the periodical table. As we go down along a column in a periodical table, the electronegativity decreases. Hence, the electron pulling effect by the X through s -induction decreases. Therefore, the polarity of the H-O bond decreases along a column on going from top to bottom.

Example:

$HOCl > HOBr > HOI$ (as decreasing order of electronegativity is Cl > Br > I)

When the element X acquire additional oxygen, the electron pulling power by the group X further increases due to increase of –I inductive effect exerted by the additional oxygen atom.

Example:

$HOCl < HOClO < HOClO_2 < HOClO_3$

It is also to note the number of oxo group operation of the number of O - H bonds

Example:

H_3PO_3 is more acidic than that of H_3PO_4

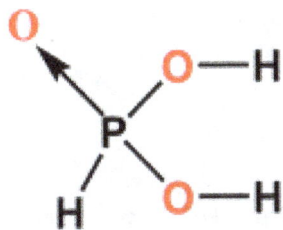 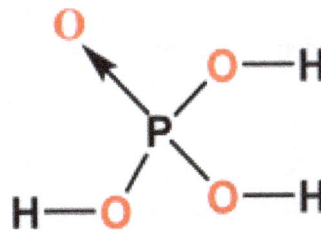

$$H_3PO_3 \qquad\qquad H_3PO_4$$

In case of H_3PO_3 one oxo group affect on two O - H bonds, while in H_3PO_4 one oxo group affect on three O - H bonds. Therefore, because of more −I inductive effect experienced by the O - H bonds present in H_3PO_3, H_3PO_4 behaves as less acidic.

Acid base behavior of metal compounds:

Metal oxides are generally basic in nature. Few of them are amphoteric. In the periodic table oxides becomes more basic on moving from top to bottom of the table, and more acidic on going from left to right.

Example:

BeO is amphoteric, MgO, CaO, SrO, BaO are basic and basicity increases from BaO to MgO

Na_2O, MgO are basic, Al_2O_3 is amphoteric, SiO_2, P_4O_{10}, SO_3, Cl_2O_7 are acidic and the acidity increases from SiO_2 to Cl_2O_7

With an increase in charge/radious ratio the acidity of metal oxides increases.

Non aqueous solvents and ionic liquids:

Studies on non-aqueous system, particularly on liquid ammonia revealed that auto-ionization of liquid NH_3 similar to water occurs.

$$NH_3 + NH_3 \leftrightarrow NH_4^+ + NH_2^-$$

Akin to substances producing H_3O^+ ions in water acts as acids in aqueous medium, substances that produce NH_4^+ ions in liquid ammonia medium acts as acids and it may also be expected to behave as an acid in that medium. Similarly, NH_2^- producing substance behaves as a base in liquid ammonia medium.

Acids in liquid ammonia medium: NH_4Cl, NH_4NO_3 , $(NH_4)_2SO_4$, etc.

Bases in liquid ammonia medium: KNH_2 , $NaNH_2$, etc.

Similarly, $SOCl_2$ behaves as an acid and $CaSO_3$ behaves as a base in liquid SO_2 solvent and their neutralization reaction can be expressed as

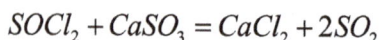

$$SOCl_2 + CaSO_3 = CaCl_2 + 2SO_2$$

Acid–base behavior is also known to show by some other ionic pair (salts). e.g. In BrF_3 , BrF_2AsF_6 increases the concentration of the cation and hence the salt is defined as an acid in the solvent system.

References

- Hall, Norris F. (March 1940). "Systems of Acids and Bases". Journal of Chemical Education. 17 (3): 124–128. Bibcode:1940JChEd..17..124H. doi:10.1021/ed017p124

- Greenwood, Norman N.; Earnshaw, Alan (1984). Chemistry of the Elements. Oxford: Pergamon Press. p. 1056. ISBN 0-08-022057-6

- Reich, Hans J. "Bordwell pKa Table (Acidity in DMSO)". Department of Chemistry, University of Wisconsin, U.S. Archived from the original on 9 October 2008. Retrieved 2008-11-02

- Germann, Albert F. O. (6 October 1925). "A General Theory of Solvent Systems". Journal of the American Chemical Society. 47 (10): 2461–2468. doi:10.1021/ja01687a006

- Flood, H.; Forland, T. (1947). "The Acidic and Basic Properties of Oxides". Acta Chemica Scandinavica. 1 (6): 592–604. PMID 18907702. doi:10.3891/acta.chem.scand.01-0592

- Masterton, William; Hurley, Cecile; Neth, Edward (2011). Chemistry: Principles and Reactions. Cengage Learning. p. 433. ISBN 1-133-38694-6

- Pearson, Ralph G. (1963). "Hard and Soft Acids and Bases". J. Am. Chem. Soc. 85 (22): 3533–3539. doi:10.1021/ja00905a001

- Lowry, T. M. (1923). "The uniqueness of hydrogen". Journal of the Society of Chemical Industry. 42 (3): 43–47. doi:10.1002/jctb.5000420302

- Ebbing, Darrell; Gammon, Steven D. (2010). General Chemistry, Enhanced Edition. Cengage Learning. pp. 644–645. ISBN 0-538-49752-1

- Cramer, R. E., and Bopp, T. T. (1977) The Great E & C Plot. A graphical display of the enthalpies of adduct formation for Lewis acids and bases.. Journal of Chemical Education 54 612-613

- "Oxonium Ylides". IUPAC Compendium of Chemical Terminology (interactive version) (2.3.3 ed.). International Union of Pure and Applied Chemistry. 2014-02-24. Retrieved 9 May 2007

- Christian Laurence and Jean-François Gal "Lewis Basicity and Affinity Scales : Data and Measurement" Wiley, 2009. ISBN 978-0-470-74957-9

- Brönsted, J. N. (1923). "Einige Bemerkungen über den Begriff der Säuren und Basen" [Some observations about the concept of acids and bases]. Recueil des Travaux Chimiques des Pays-Bas. 42 (8): 718–728. doi:10.1002/recl.19230420815

Reduction–Oxidation Reactions

A reaction where the oxidation states of atoms are changed are known as reduction-oxidation reactions or redox reactions. Redox potential is the discerning of the driving force of the reaction as being either reductive or a change in states. The chapter on reduction-oxidation reactions offers an insightful focus, keeping in mind the complex subject matter.

Redox

The two parts of a redox reaction

Reduction

Oxidant + e⁻ ⟶ Product

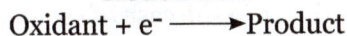

(Gain of Electrons) (Oxidation Number Decreases)

Oxidation

Reductant ⟶ Product + e⁻

(Loss of Electrons) (Oxidation Number Increases)

Rust, a slow redox reaction

Redox (short for reduction–oxidation reaction) is a chemical reaction in which the oxidation states of atoms are changed. Any such reaction involves both a reduction process and a complementary oxidation process, two key concepts involved with electron transfer processes. Redox reactions include all chemical reactions in which atoms have their oxidation state changed; in general, redox reactions involve the transfer of electrons between chemical species. The chemical species from which the electron is stripped is said to have been oxidized, while the chemical species to which the electron is added is said to have been reduced. It can be explained in simple terms:

- Oxidation is the loss of electrons or an increase in oxidation state by a molecule, atom, or ion.

- Reduction is the gain of electrons or a *decrease* in oxidation state by a molecule, atom, or ion.

As an example, during the combustion of wood, oxygen from the air is reduced, gaining electrons from the carbon. Although oxidation reactions are commonly associated with the formation of oxides from oxygen molecules, oxygen is not necessarily included in such reactions, as other chemical species can serve the same function.

A bonfire; combustion is a fast redox reaction

The reaction can occur relatively slowly, as in the case of rust, or more quickly, as in the case of fire. There are simple redox processes, such as the oxidation of carbon to yield carbon dioxide (CO_2) or the reduction of carbon by hydrogen to yield methane (CH_4), and more complex processes such as the oxidation of glucose ($C_6H_{12}O_6$) in the human body.

Etymology

"Redox" is a portmanteau of "reduction" and "oxidation".

The word *oxidation* originally implied reaction with oxygen to form an oxide, since dioxygen (O_2 (g)) was historically the first recognized oxidizing agent. Later, the term was expanded to encompass oxygen-like substances that accomplished parallel chemical reactions. Ultimately, the meaning was generalized to include all processes involving loss of electrons.

The word *reduction* originally referred to the loss in weight upon heating a metallic ore such as a metal oxide to extract the metal. In other words, ore was "reduced" to metal. Antoine Lavoisier (1743–1794) showed that this loss of weight was due to the loss of oxygen as a gas. Later, scientists realized that the metal atom gains electrons in this process. The meaning of *reduction* then became generalized to include all processes involving gain of electrons. Even though "reduction" seems counter-intuitive when speaking of the gain of electrons, it might help to think of reduction as the loss of oxygen, which was its historical meaning. Since electrons are negatively charged, it is also helpful to think of this as reduction in electrical charge.

The electrochemist John Bockris has used the words *electronation* and *deelectronation* to describe reduction and oxidation processes respectively when they occur at electrodes. These words are analogous to protonation and deprotonation, but they have not been widely adopted by chemists.

The term "hydrogenation" could be used instead of reduction, since hydrogen is the reducing agent in a large number of reactions, especially in organic chemistry and biochemistry. But, unlike oxidation, which has been generalized beyond its root element, hydrogenation has maintained its specific connection to reactions that *add* hydrogen to another substance (e.g., the hydrogenation of unsaturated fats into saturated fats, $R-CH=CH-R + H_2 \rightarrow R-CH_2-CH_2-R$). The word "redox" was first used in 1928.

Definitions

The processes of oxidation and reduction occur simultaneously and cannot happen independently of one another, similar to the acid–base reaction. The oxidation alone and the reduction alone are each called a *half-reaction*, because two half-reactions always occur together to form a whole reaction. When writing half-reactions, the gained or lost electrons are typically included explicitly in order that the half-reaction be balanced with respect to electric charge.

Though sufficient for many purposes, these general descriptions are not precisely correct. Although oxidation and reduction properly refer to *a change in oxidation state* — the actual transfer of electrons may never occur. The oxidation state of an atom is the fictitious charge that an atom would have if all bonds between atoms of different elements were 100% ionic. Thus, oxidation is best defined as an *increase in oxidation state*, and reduction as a *decrease in oxidation state*. In practice, the transfer of electrons will always cause a change in oxidation state, but there are many reactions that are classed as "redox" even though no electron transfer occurs (such as those involving covalent bonds).

Oxidizing and Reducing Agents

In redox processes, the reductant transfers electrons to the oxidant. Thus, in the reaction, the reductant or *reducing agent* loses electrons and is oxidized, and the oxidant or *oxidizing agent* gains electrons and is reduced. The pair of an oxidizing and reducing agent that are involved in a particular reaction is called a redox pair. A redox couple is a reducing species and its corresponding oxidizing form, e.g., Fe^{2+}/Fe^{3+}.

Oxidizers

The international pictogram for oxidizing chemicals.

Substances that have the ability to oxidize other substances (cause them to lose electrons) are said to be oxidative or oxidizing and are known as oxidizing agents, oxidants, or oxidizers. That is, the oxidant (oxidizing agent) removes electrons from another substance, and is thus itself reduced. And, because it "accepts" electrons, the oxidizing agent is also called an electron acceptor. Oxygen is the quintessential oxidizer.

Oxidants are usually chemical substances with elements in high oxidation states (e.g., H_2O_2, MnO_4^-, CrO_3, $Cr_2O_7^{2-}$, OsO_4), or else highly electronegative elements (O_2, F_2, Cl_2, Br_2) that can gain extra electrons by oxidizing another substance.

Reducers

Substances that have the ability to reduce other substances (cause them to gain electrons) are said to be reductive or reducing and are known as reducing agents, reductants, or reducers. The reductant (reducing agent) transfers electrons to another substance, and is thus itself oxidized. And, because it "donates" electrons, the reducing agent is also called an electron donor. Electron donors can also form charge transfer complexes with electron acceptors.

Reductants in chemistry are very diverse. Electropositive elemental metals, such as lithium, sodium, magnesium, iron, zinc, and aluminium, are good reducing agents. These metals donate or *give away* electrons readily. *Hydride transfer reagents*, such as $NaBH_4$ and $LiAlH_4$, are widely used in organic chemistry, primarily in the reduction of carbonyl compounds to alcohols. Another method of reduction involves the use of hydrogen gas (H_2) with a palladium, platinum, or nickel catalyst. These *catalytic reductions* are used primarily in the reduction of carbon-carbon double or triple bonds.

Standard Electrode Potentials (Reduction Potentials)

Each half-reaction has a *standard electrode potential* (E_{cell}^0), which is equal to the potential difference or voltage at equilibrium under standard conditions of an electrochemical cell in which the cathode reaction is the half-reaction considered, and the anode is a standard hydrogen electrode where hydrogen is oxidized:

$$\tfrac{1}{2} H_2 \rightarrow H^+ + e^-.$$

The electrode potential of each half-reaction is also known as its *reduction potential* E_{red}^0 , or potential when the half-reaction takes place at a cathode. The reduction potential is a measure of the tendency of the oxidizing agent to be reduced. Its value is zero for $H^+ + e^- \rightarrow \tfrac{1}{2} H_2$ by definition, positive for oxidizing agents stronger than H^+ (e.g., +2.866 V for F_2) and negative for oxidizing agents that are weaker than H^+ (e.g., −0.763 V for Zn^{2+}).

For a redox reaction that takes place in a cell, the potential difference is:

$$E_{cell}^0 = E_{cathode}^0 - E_{anode}^0$$

However, the potential of the reaction at the anode was sometimes expressed as an *oxidation potential*:

$$E_{ox}^0 = -E_{red}^0$$

The oxidation potential is a measure of the tendency of the reducing agent to be oxidized, but does not represent the physical potential at an electrode. With this notation, the cell voltage equation is written with a plus sign

$$E_{cell}^0 = E_{red(cathode)}^0 + E_{ox(anode)}^0$$

Examples of Redox Reactions

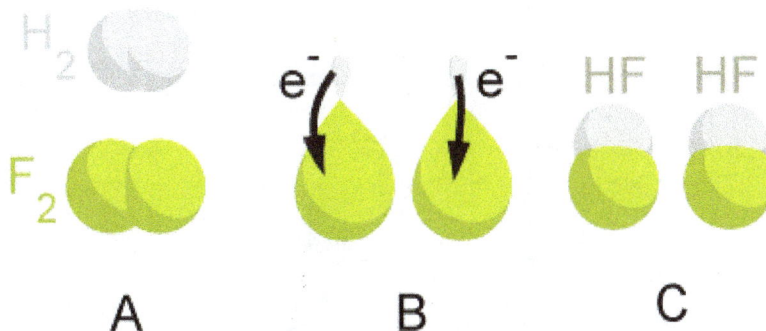

Illustration of a redox reaction

A good example is the reaction between hydrogen and fluorine in which hydrogen is being oxidized and fluorine is being reduced:

$$H_2 + F_2 \rightarrow 2\,HF$$

We can write this overall reaction as two half-reactions:

the oxidation reaction:

$$H_2 \rightarrow 2\,H^+ + 2\,e^-$$

and the reduction reaction:

$$F_2 + 2\,e^- \rightarrow 2\,F^-$$

Analyzing each half-reaction in isolation can often make the overall chemical process clearer. Because there is no net change in charge during a redox reaction, the number of electrons in excess in the oxidation reaction must equal to the number consumed by the reduction reaction.

Elements, even in molecular form, always have an oxidation state of zero. In the first half-reaction, hydrogen is oxidized from an oxidation state of zero to an oxidation state of +1. In the second half-reaction, fluorine is reduced from an oxidation state of zero to an oxidation state of −1.

When adding the reactions together the electrons are canceled:

$$H_2 \rightarrow 2\,H^+ + 2\,e^-$$

$$F_2 + 2\,e^- \rightarrow 2\,F^-$$

$$H_2 + F_2 \rightarrow 2\,H^+ + 2\,F^-$$

And the ions combine to form hydrogen fluoride:

$$2\,H^+ + 2\,F^- \rightarrow 2\,HF$$

The overall reaction is:

$$H_2 + F_2 \rightarrow 2\,HF$$

Metal Displacement

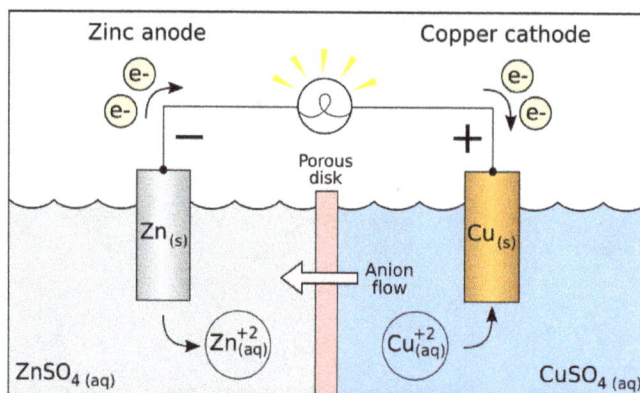

A redox reaction is the force behind an electrochemical cell like the Galvanic
cell pictured. The battery is made out of a zinc electrode in a $ZnSO_4$ solution
connected with a wire and a porous disk to a copper electrode in a $CuSO_4$ solution.

In this type of reaction, a metal atom in a compound (or in a solution) is replaced by an atom of
another metal. For example, copper is deposited when zinc metal is placed in a copper(II) sulfate
solution:

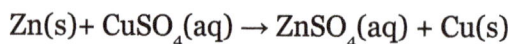

$$Zn(s) + CuSO_4(aq) \rightarrow ZnSO_4(aq) + Cu(s)$$

In the above reaction, zinc metal displaces the copper(II) ion from copper sulfate solution and thus
liberates free copper metal.

The ionic equation for this reaction is:

$$Zn + Cu^{2+} \rightarrow Zn^{2+} + Cu$$

As two half-reactions, it is seen that the zinc is oxidized:

$$Zn \rightarrow Zn^{2+} + 2\ e^-$$

And the copper is reduced:

$$Cu^{2+} + 2\ e^- \rightarrow Cu$$

Other Examples

* The reduction of nitrate to nitrogen in the presence of an acid (denitrification):

$$2\ NO_3^- + 10\ e^- + 12\ H^+ \rightarrow N_2 + 6\ H_2O$$

* The combustion of hydrocarbons, such as in an internal combustion engine, which produces water, carbon dioxide, some partially oxidized forms such as carbon monoxide, and
heat energy. Complete oxidation of materials containing carbon produces carbon dioxide.

* In organic chemistry, the stepwise oxidation of a hydrocarbon by oxygen produces water
and, successively, an alcohol, an aldehyde or a ketone, a carboxylic acid, and then a peroxide.

Corrosion and Rusting

Oxides, such as iron(III) oxide or rust, which consists of hydrated iron(III) oxides $Fe_2O_3 \cdot nH_2O$ and iron(III) oxide-hydroxide ($FeO(OH)$, $Fe(OH)_3$), form when oxygen combines with other elements

Iron rusting in pyrite cubes

- The term corrosion refers to the electrochemical oxidation of metals in reaction with an oxidant such as oxygen. Rusting, the formation of iron oxides, is a well-known example of electrochemical corrosion; it forms as a result of the oxidation of iron metal. Common rust often refers to iron(III) oxide, formed in the following chemical reaction:

$$4\,Fe + 3\,O_2 \rightarrow 2\,Fe_2O_3$$

- The oxidation of iron(II) to iron(III) by hydrogen peroxide in the presence of an acid:

$$Fe^{2+} \rightarrow Fe^{3+} + e^-$$

$$H_2O_2 + 2\,e^- \rightarrow 2\,OH^-$$

Overall equation:

$$2\,Fe^{2+} + H_2O_2 + 2\,H^+ \rightarrow 2\,Fe^{3+} + 2\,H_2O$$

Redox Reactions in Industry

Cathodic protection is a technique used to control the corrosion of a metal surface by making it the cathode of an electrochemical cell. A simple method of protection connects protected metal to a more easily corroded "sacrificial anode" to act as the anode. The sacrificial metal instead of the protected metal, then, corrodes. A common application of cathodic protection is in galvanized steel, in which a sacrificial coating of zinc on steel parts protects them from rust.

The primary process of reducing ore at high temperature to produce metals is known as smelting.

Oxidation is used in a wide variety of industries such as in the production of cleaning products and oxidizing ammonia to produce nitric acid, which is used in most fertilizers.

Redox reactions are the foundation of electrochemical cells, which can generate electrical energy or support electrosynthesis.

The process of electroplating uses redox reactions to coat objects with a thin layer of a material, as in chrome-plated automotive parts, silver plating cutlery, and gold-plated jewelry.

The production of compact discs depends on a redox reaction, which coats the disc with a thin layer of metal film.

Redox Reactions in Biology

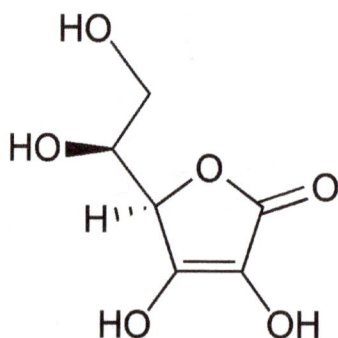

ascorbic acid (reduced form of Vitamin C)

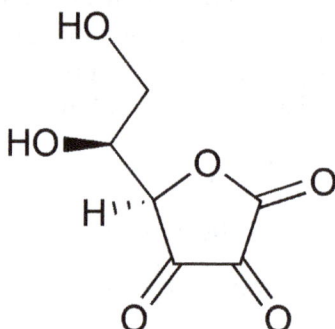

dehydroascorbic acid (oxidized form of Vitamin C)

Many important biological processes involve redox reactions.

Cellular respiration, for instance, is the oxidation of glucose ($C_6H_{12}O_6$) to CO_2 and the reduction of oxygen to water. The summary equation for cell respiration is:

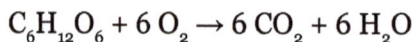

$$C_6H_{12}O_6 + 6\,O_2 \rightarrow 6\,CO_2 + 6\,H_2O$$

The process of cell respiration also depends heavily on the reduction of NAD^+ to NADH and the reverse reaction (the oxidation of NADH to NAD^+). Photosynthesis and cellular respiration are complementary, but photosynthesis is not the reverse of the redox reaction in cell respiration:

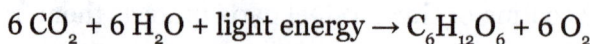

$$6\,CO_2 + 6\,H_2O + \text{light energy} \rightarrow C_6H_{12}O_6 + 6\,O_2$$

Biological energy is frequently stored and released by means of redox reactions. Photosynthesis involves the reduction of carbon dioxide into sugars and the oxidation of water into molecular oxygen. The reverse reaction, respiration, oxidizes sugars to produce carbon dioxide and water. As intermediate steps, the reduced carbon compounds are used to reduce nicotinamide adenine dinucleotide (NAD^+), which then contributes to the creation of a proton gradient, which drives the synthesis of adenosine triphosphate (ATP) and is maintained by the reduction of oxygen. In animal cells, mitochondria perform similar functions.

Free radical reactions are redox reactions that occur as a part of homeostasis and killing microorganisms, where an electron detaches from a molecule and then reattaches almost instantaneously. Free radicals are a part of redox molecules and can become harmful to the human body if they do not reattach to the redox molecule or an antioxidant. Unsatisfied free radicals can spur the mutation of cells they encounter and are, thus, causes of cancer.

The term redox state is often used to describe the balance of GSH/GSSG, NAD^+/NADH and $NADP^+$/ NADPH in a biological system such as a cell or organ. The redox state is reflected in the balance of several sets of metabolites (e.g., lactate and pyruvate, beta-hydroxybutyrate, and acetoacetate), whose interconversion is dependent on these ratios. An abnormal redox state can develop in a variety of deleterious situations, such as hypoxia, shock, and sepsis. Redox mechanism also control some cellular processes. Redox proteins and their genes must be co-located for redox regulation according to the CoRR hypothesis for the function of DNA in mitochondria and chloroplasts.

Redox Cycling

A wide variety of aromatic compounds are enzymatically reduced to form free radicals that contain one more electron than their parent compounds. In general, the electron donor is any of a wide variety of flavoenzymes and their coenzymes. Once formed, these anion free radicals reduce molecular oxygen to superoxide, and regenerate the unchanged parent compound. The net reaction is the oxidation of the flavoenzyme's coenzymes and the reduction of molecular oxygen to form superoxide. This catalytic behavior has been described as futile cycle or redox cycling.

Examples of redox cycling-inducing molecules are the herbicide paraquat and other viologens and quinones such as menadione.

Redox Reactions in Geology

Mi Vida uranium mine, near Moab, Utah. The alternating red and white/green bands of sandstone correspond to oxidized and reduced conditions in groundwater redox chemistry.

In geology, redox is important to both the formation of minerals and the mobilization of minerals, and is also important in some depositional environments. In general, the redox state of most rocks can be seen in the color of the rock. The rock forms in oxidizing conditions, giving it a red color. It is then "bleached" to a green—or sometimes white—form when a reducing fluid passes through the rock. The reduced fluid can also carry uranium-bearing minerals. Famous examples of redox conditions affecting geological processes include uranium deposits and Moqui marbles.

Balancing Redox Reactions

Describing the overall electrochemical reaction for a redox process requires a *balancing* of the component half-reactions for oxidation and reduction. In general, for reactions in aqueous solution, this involves adding H^+, OH^-, H_2O, and electrons to compensate for the oxidation changes.

Acidic Media

In acidic media, H^+ ions and water are added to half-reactions to balance the overall reaction.

For instance, when manganese(II) reacts with sodium bismuthate:

Unbalanced reaction:	$Mn^{2+}(aq) + NaBiO_3(s) \rightarrow Bi^{3+}(aq) + MnO_4^-(aq)$
Oxidation:	$4\,H_2O(l) + Mn^{2+}(aq) \rightarrow MnO_4^-(aq) + 8\,H^+(aq) + 5\,e^-$
Reduction:	$2\,e^- + 6\,H^+ + BiO_3^-(s) \rightarrow Bi^{3+}(aq) + 3\,H_2O(l)$

The reaction is balanced by scaling the two half-cell reactions to involve the same number of electrons (multiplying the oxidation reaction by the number of electrons in the reduction step and vice versa):

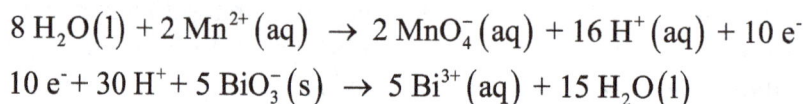

$$8\,H_2O(l) + 2\,Mn^{2+}(aq) \rightarrow 2\,MnO_4^-(aq) + 16\,H^+(aq) + 10\,e^-$$
$$10\,e^- + 30\,H^+ + 5\,BiO_3^-(s) \rightarrow 5\,Bi^{3+}(aq) + 15\,H_2O(l)$$

Adding these two reactions eliminates the electrons terms and yields the balanced reaction:

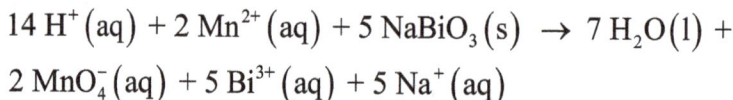

$$14\,H^+(aq) + 2\,Mn^{2+}(aq) + 5\,NaBiO_3(s) \rightarrow 7\,H_2O(l) +$$
$$2\,MnO_4^-(aq) + 5\,Bi^{3+}(aq) + 5\,Na^+(aq)$$

Basic Media

In basic media, OH^- ions and water are added to half reactions to balance the overall reaction.

For example, in the reaction between potassium permanganate and sodium sulfite:

Unbalanced reaction:	$KMnO_4 + Na_2SO_3 + H_2O \rightarrow MnO_2 + Na_2SO_4 + KOH$
Reduction:	$3\,e^- + 2\,H_2O + MnO_4^- \rightarrow MnO_2 + 4\,OH^-$
Oxidation:	$2\,OH^- + SO_3^{2-} \rightarrow SO_4^{2-} + H_2O + 2\,e^-$

Balancing the number of electrons in the two half-cell reactions gives:

$$6\,e^- + 4\,H_2O + 2\,MnO_4^- \rightarrow 2\,MnO_2 + 8\,OH^-$$

$$6\,OH^- + 3\,SO_3^{2-} \rightarrow 3\,SO_4^{2-} + 3\,H_2O + 6\,e^-$$

Adding these two half-cell reactions together gives the balanced equation:

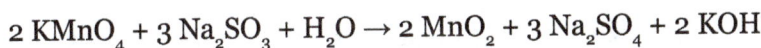

$$2\,KMnO_4 + 3\,Na_2SO_3 + H_2O \rightarrow 2\,MnO_2 + 3\,Na_2SO_4 + 2\,KOH$$

Memory Aids

The key terms involved in redox are often confusing. For example, a reagent that is oxidized loses electrons; however, that reagent is referred to as the reducing agent. Likewise, a reagent that is reduced gains electrons and is referred to as the oxidizing agent. Acronyms or mnemonics are commonly used to help remember the terminology:

- "OIL RIG" — oxidation is loss of electrons, reduction is gain of electrons.

- "LEO the lion says GER" — loss of electrons is oxidation, gain of electrons is reduction.

- "LEORA says GEROA" — loss of electrons is oxidation (reducing agent), gain of electrons is reduction (oxidizing agent).

- "RED CAT" and "AN OX", or "AnOx RedCat" ("an ox-red cat") — reduction occurs at the cathode and the anode is for oxidation.

- "RED CAT gains what AN OX loses" – reduction at the cathode gains (electrons) what anode oxidation loses (electrons).

Redox Potential

The reactions which involve oxidation & reduction are called redox reaction.

e.g.

$$\overset{\text{Reduction}}{4HCl + MnO_2} = MnCl_2 + Cl_2 + 2H_2O$$

Oxidation

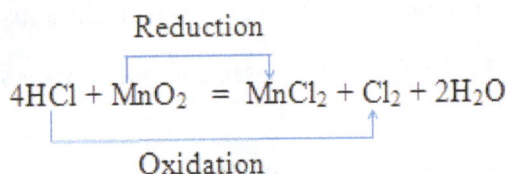

In here HCl has been oxidised to Cl_2 and MnO_2 has been reduced to $MnCl_2$.

Oxidation number : It's the charge which an atom of an element has in its ion or appear to have when present in the combined state. It is also known as oxidation state.

Rules for calculation of oxidation number:

1. Oxidation number of any atom in the elementary state is zero.

   e.g. in O_2, H_2, Na, He and Fe the oxidation state of each atom equal to zero.

2. Oxidation number of mono atomic ion is equal to the charge on it.

3. Oxidation number of H is +1 when combined with non metal and -1 when combined with active metal like Na, Ca etc. e.g. NaH, CaH_2.

4. Oxidation number of oxygen is -2 except in peroxides like H_2O_2, Na_2O_2 etc where it is -1 and OF_2 where it is +2.

5. Oxidation number of alkali and alkaline earth metals is +1 and +2 respectively.

6. Oxidation number of halogens is in -1 in metal halides.

7. In compounds of metal with non-metal, metal have positive oxidation numbers whereas non-metals have negative Oxidation numbers.

8. In compound of two different elements, the more electronegative has negative Oxidation number and the other has positive oxidation number.

9. In complex ions, the sum of the oxidation numbers of all atoms is equal to the charge on the ion.

Oxidation (de-electronation): loss of electron or result in the increase in oxidation number of its atom/s.

Oxidising agent: acceptor of electron/s.

Reduction (Electronation): Gain of electron/s or decrease in oxidation number of its atom/s.

Reducing agent: Donor of electron/s.

Any oxidation -reduction (redox) reaction can be divided into two half reactions: one in which a chemical species undergoes oxidation and one in which another chemical species undergoes reduction. If a half- reaction is written as a reduction, the driving force is the reduction potential. If the half-reaction is written as oxidation , the driving force is the oxidation potential related to the reduction potential by a sign change. So the redox potential is the reduction/ oxidation potential of a compound measured under standards conditions against a standard reference half-cell.

In biological systems the standard redox potential is defined at pH – 7.0 versus the hydrogen electrode and partial pressure of hydrogen = 1 bar .

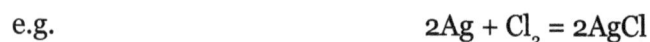

e.g. $2Ag + Cl_2 = 2AgCl$

The above redox reaction can be split into two half reactions

$2\ Ag = 2Ag^+ + 2e$ or $Ag = Ag^+ + e$ - (I)

$Cl_2 + 2e = 2Cl^-$ or $½\ Cl_2 + e = Cl^-$ - (II)

Reaction (I) is a oxidation process & the potential of this reaction is oxidation potential whereas potential of reaction (II) is reduction potential.

Analysis of redox cycle:

The half-cell potential of two half cells is given below. What is the redox reaction that takes place when they are combined in a favourable condition?

$$E^0_{Fe^{3+},Fe^{2+}|PT} = 0.177 \text{ V and } E^0_{MnO_4^-,Mn^{2+},H^+|Pt} = 1.51v$$

Hints: $Pt|Fe^{3+},Fe^{2+}||MnO_4^-,Mn^{2+},H^+|Pt$

Reaction happened

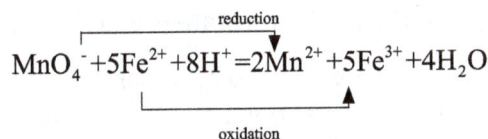

$$\text{MnO}_4^- + 5\text{Fe}^{2+} + 8\text{H}^+ = 2\text{Mn}^{2+} + 5\text{Fe}^{3+} + 4\text{H}_2\text{O}$$

reduction

oxidation

Oxidation State

The oxidation state, often called the oxidation number, is an indicator of the degree of oxidation (loss of electrons) of an atom in a chemical compound. Conceptually, the oxidation state, which may be positive, negative or zero, is the hypothetical charge that an atom would have if all bonds to atoms of different elements were 100% ionic, with no covalent component. This is never exactly true for real bonds. High oxidation numbers generally correlate with high valences, although the concepts are not equivalent.

The term "oxidation" was first used by Antoine Lavoisier to signify reaction of a substance with oxygen. Much later, it was realized that the substance, upon being oxidized, loses electrons, and the use of the term "oxidation" was extended to include other reactions in which electrons are lost.

Oxidation states are typically represented by integers. In some cases, the average oxidation state of an element is a fraction, such as $\frac{8}{3}$ for iron in magnetite (Fe_3O_4). The highest known oxidation state is reported to be +9 in the iridium tetroxide cation (IrO_4^+), while the lowest known oxidation state is −5 for boron, gallium, indium, and thallium in various Zintl phases, a type of intermetallic compound. It is predicted that even a +10 oxidation state may be achieveable by platinum in the platinum tetroxide dication (PtO_4^{2+}).

The increase in oxidation state of an atom, through a chemical reaction, is known as an oxidation; a decrease in oxidation state is known as a reduction. Such reactions involve the formal transfer of electrons: a net gain in electrons being a reduction, and a net loss of electrons being an oxidation. For pure elements, the oxidation state is zero.

There are various methods for determining oxidation states.

In inorganic nomenclature, the oxidation state is determined and expressed as an oxidation number, and is represented by a Roman numeral placed after the element name.

In coordination chemistry, "oxidation number" is defined differently from "oxidation state".

IUPAC Definitions of Oxidation State and Oxidation Number

Oxidation State

A "Comprehensive definition of oxidation state (IUPAC Recommendations 2016)" has been published with free access. This is a distillation of an IUPAC technical report "Toward a comprehensive definition of oxidation state". The current *Gold Book* definition of oxidation state listed by IUPAC is as follows:

"Oxidation state" is defined as the charge an atom might be imagined to have when electrons are counted according to an agreed-upon set of rules:

1. the oxidation state of a free element (uncombined element) is zero

2. for a simple (monatomic) ion, the oxidation state is equal to the net charge on the ion

3. hydrogen has an oxidation state of +1 and oxygen has an oxidation state of −2 when they are present in most compounds. Exceptions to this are that hydrogen has an oxidation state of −1 in hydrides of active metals, e.g. LiH, and oxygen has an oxidation state of −1 in peroxides, e.g. H_2O_2.

4. the algebraic sum of oxidation states of all atoms in a neutral molecule must be zero, while in ions the algebraic sum of the oxidation states of the constituent atoms must be equal to the charge on the ion.

Determining the Oxidation State or Number

There are two different methods for determining the oxidation state of elements in chemical compounds. First (and widely taught) a method based on the rules in the IUPAC definition. Second, a method based on the relative electronegativity of the elements in the compound, where the more electronegative element is assumed to take the negative charge.

Simple Examples using IUPAC Definition

- Any pure element—even if it forms diatomic molecules like chlorine (Cl_2)—has an oxidation state of zero. Examples of this are Cu or O_2.

- For monatomic ions, the oxidation state is the same as the charge of the ion. For example, the sulfide anion (S^{2-}) has an oxidation state of −2, whereas the lithium cation (Li^+) has an oxidation state of +1.

- The sum of oxidation states for all atoms in a molecule or polyatomic ion is equal to the charge of the molecule or ion. Thus, the oxidation state of one element can be calculated from the oxidation states of the other elements.

  1. An application of this rule is that the sum of the oxidation states of all atoms in a neutral molecule must be zero. Consider a neutral molecule of carbon dioxide, CO_2. Oxygen is assumed to have its usual oxidation state of −2, and so the sum of the oxidation states of all the atoms can be expressed as $x + 2(-2) = 0$, or $x - 4 = 0$, where x is the unknown oxidation state of carbon. Thus, it can be seen that the oxidation state of carbon in the molecule is +4.

  2. In polyatomic ions, the sum of the oxidation states of the constituent atoms must be equal to the charge on the ion. As an example, consider the sulfate anion, which has the formula SO_4^{2-}. As indicated by the formula, the total charge of this ion is −2. Because all four oxygen atoms are assumed to have their usual oxidation state of −2, and the

sum of the oxidation states of all the atoms is equal to the charge of the ion, the sum of the oxidation states can be represented as $y + 4(-2) = -2$, or $y - 8 = -2$, where y is the unknown oxidation state of sulfur. Thus, it can be computed that $y = +6$.

These facts, combined with some elements almost always having certain oxidation states (due to their very high electropositivity or electronegativity), allows one to compute the oxidation states for the remaining atoms (such as transition metals) in simple compounds.

Example for a complex salt: In $Cr(OH)_3$, oxygen has an oxidation state of -2 (no fluorine or O–O bonds present), and hydrogen has a state of $+1$ (bonded to oxygen). So, each of the three hydroxide groups has an overall oxidation state of $-2 + 1 = -1$. As the compound is neutral, chromium has an oxidation state of $+3$.

Using Electronegativity

The use of electronegativity in this way was introduced by Linus Pauling in 1947. This method of determining oxidation state is found in some recent text books. This method allows the oxidation state of all atoms in a molecule to be determined whereas the IUPAC 1990/2005 definition does not. In the 1970 rules, IUPAC recommended that oxidation state be used in nomenclature and elsewhere in inorganic chemistry as the "charge that would be present on an atom if the electrons were assigned to the more electronegative atom", but with a convention that hydrogen is considered to be positive in combination with nonmetals and a bond between like atoms makes no contribution to the oxidation number.

In practice the IUPAC 1990/2005 definition is usually extended by adding additional rules based on electronegativity.

- Fluorine has an oxidation state of -1 when bonded to any other element, since it has the highest electronegativity of all reactive elements.

- Halogens other than fluorine have an oxidation state of -1 except when they are bonded to oxygen, to nitrogen, or to another halogen that is more electronegative. For example, the oxidation state of chlorine in chlorine monofluoride (ClF) is $+1$. However, in bromine monochloride (BrCl), the oxidation state of Cl is -1.

- Hydrogen has an oxidation state of $+1$ except when bonded to more electropositive elements such as sodium, aluminium, and boron, as in NaH, $NaBH_4$, $LiAlH_4$, where each H has an oxidation state of -1.

- In compounds, oxygen typically has an oxidation state of -2, though there are exceptions that are listed below, such as peroxides (e.g. hydrogen peroxide H_2O_2), where oxygen has an oxidation state of -1.

- Alkali metals have an oxidation state of $+1$ in virtually all of their compounds.

- Alkaline earth metals have an oxidation state of $+2$ in virtually all of their compounds.

Ionic Compounds

If an ionic compound has two ions with a common element, such as ammonium nitrate, NH_4NO_3, it is usual to consider the oxidation states for each ion separately. According to that empirical formula, one can use the standard charges of hydrogen and oxygen to solve for the charge of nitrogen: $2x + 4(1) + 3(-2) = 0$, giving an average nitrogen oxidation state $x = +1$. However it is more correct to consider separately the ions NH_4^+ and NO_3^- with nitrogen oxidation states of -3 and $+5$ respectively, because the two nitrogen atoms are in different covalent environments.

Calculation of Oxidation States with a Lewis Structure

This method can be used for molecules when one has a Lewis structure.

It should be remembered that the oxidation state of an atom does not represent the "real" charge on that atom. This is particularly true of high oxidation states, where the ionization energy required to produce a multiply positive ion are far greater than the energies available in chemical reactions. The assignment of electrons between atoms in calculating an oxidation state is purely a formalism, but is a useful one for the understanding of many chemical reactions.

The Lewis Structure

When a Lewis structure of a molecule is available, the oxidation states may be assigned by computing the difference between the number of valence electrons that a neutral atom of that element would have and the number of electrons that "belong" to it in the Lewis structure. For purposes of computing oxidation states, electrons in a bond between atoms of different elements belong to the more electronegative atom; electrons in a bond between atoms of the same element are split equally, and electrons in a lone pair belong only to the atom with the lone pair.

For example, consider acetic acid:

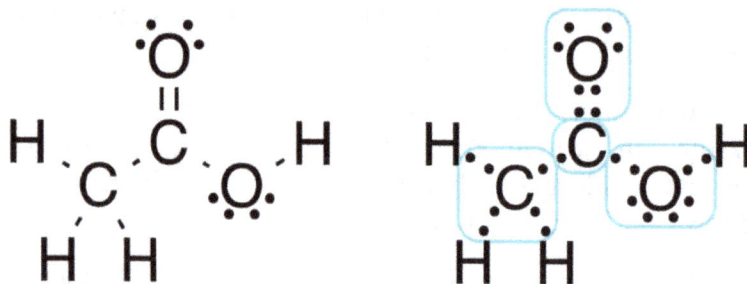

The methyl group carbon atom has six valence electrons from its bonds to the hydrogen atoms because carbon is more electronegative than hydrogen. Also, one electron is gained from its bond with the other carbon atom because the electron pair in the C–C bond is split equally. It therefore has a total of seven electrons, whereas a neutral carbon atom would have four valence electrons because carbon is in group 14 of the periodic table. The difference, $4 - 7 = -3$, is the oxidation state of that carbon atom. That is, if it is assumed that all the bonds were 100% ionic (which in fact they are not), the carbon would be described as C^{3-}.

Following the same rules, the carboxylic acid carbon atom gets the other one valence electron from the C–C bond. None of its other valence electrons are counted here because they are involved in bonding to oxygen atoms, which are more electronegative than carbon. For a carbon with one electron, the oxidation state is +3, again calculated against its neutral four-electron count. The oxygen atoms each have an oxidation state of −2; they get 8 electrons each (4 from the lone pairs and 4 from the bonds), while a neutral oxygen atom would have 6. The hydrogen atoms each have oxidation state +1, because they surrender their electron to the more electronegative atoms to which they are bonded.

In structural diagrams for organic chemistry, oxidation states are represented by Roman numerals to distinguish them from formal charges (calculated with all bonds covalent).

Inequivalent Atoms of an Element

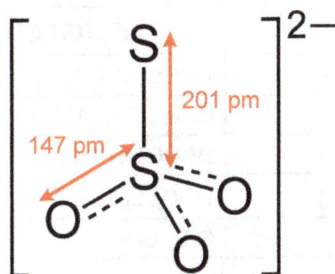

Structure of the thiosulfate anion

An example of a molecule with inequivalent atoms of the same element is the thiosulfate ion ($S_2O_3^{2-}$), for which the algebraic sum rule yields the average value +2 for sulfur, where the two ionizing electrons are assigned to the terminal sulfur atom. However, the use of a Lewis structure and electron counting shows that the two sulfur atoms are different. The central sulfur is assigned only one valence electron from the S–S bond and no valence electrons from the S–O bonds, compared to six valence electrons for a free sulfur atom, so the oxidation state of the central sulfur is +5. The terminal sulfur atom is assigned the other electron from the S–S bond plus three lone pairs for a total of seven valence electrons, so its oxidation state is −1.

Redox Reactions

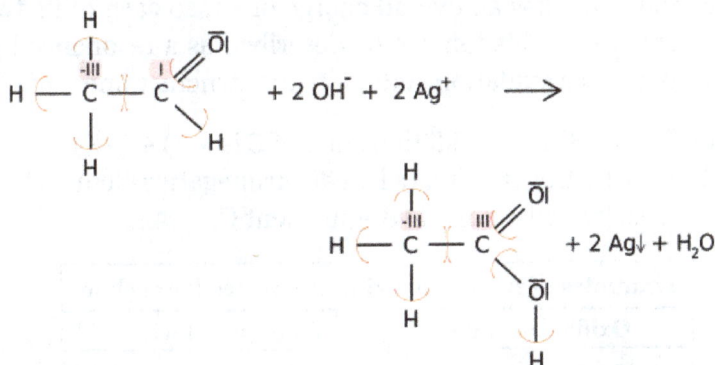

Oxidation states can be useful for balancing chemical equations for oxidation–reduction (or redox) reactions, because the changes in the oxidized atoms have to be balanced by the changes in the reduced atoms. For example, in the reaction of acetaldehyde with the Tollens' reagent to acetic acid, the carbonyl carbon atom changes its oxidation state from +1 to +3 (oxidation). This oxidation is balanced by reducing two equivalents of silver from Ag^+ to Ag^0.

Elements with Multiple Oxidation States

Most elements have more than one possible oxidation state. For example, carbon has nine integer oxidation states:

Integer oxidation states of carbon	
Oxidation state	Example compound
−4	CH_4
−3	C_2H_6
−2	C_2H_4, CH_3Cl
−1	C_2H_2, C_6H_6, $(CH_2OH)_2$
0	HCHO, CH_2Cl_2
+1	OCHCHO, $CHCl_2CHCl_2$
+2	HCOOH, $CHCl_3$
+3	HOOCCOOH, C_2Cl_6
+4	CCl_4, CO_2

Fractional Oxidation States

Fractional oxidation states are often used to represent the average oxidation states of several atoms of the same element in a structure. For example, the formula of magnetite is Fe_3O_4, implying an oxidation state for iron of $+8/3$. However this average value may not be representative if the atoms are not equivalent. In an Fe_3O_4 crystal, two-thirds of the iron ions are Fe^{3+} and one-third Fe^{2+}, and the formula may be better represented as $FeO \cdot Fe_2O_3$.

Likewise, propane, C_3H_8, has been described as having a carbon oxidation state of $-8/3$. Again this is an average value since the structure of the molecule is $H_3C-CH_2-CH_3$, with the first and third carbon atoms each having an oxidation state of −3 and the central one −2.

An example with true fractional oxidation states for equivalent atoms is potassium superoxide, KO_2. The diatomic superoxide ion has an overall charge of −1, so each of its two oxygen atoms is assigned an oxidation state of $-1/2$. This ion can be described as a resonance hybrid of two Lewis structures, where each oxygen has oxidation state 0 in one structure and −1 in the other.

For the cyclopentadienyl ion $C_5H_5^-$, the oxidation state of C is $(-1) + (-1/5) = -6/5$. The −1 occurs because each C is bonded to one hydrogen atom (a less electronegative element), and the $-1/5$ because the total ionic charge of −1 is divided among five equivalent C.

Examples of fractional oxidation states for carbon	
Oxidation state	Example species
$-6/5$	$C_5H_5^-$
$-6/7$	$C_7H_7^+$
$+3/2$	$C_4O_4^{2-}$

Oxidation State and Formal Charge

The oxidation state of an atom is often different from the formal charge often included in Lewis structures (when it is non-zero). The oxidation state is calculated by assuming that each chemical bond (except between identical atoms) is ionic so that both electrons are assigned to the more electronegative bonded atom. In contrast, the formal charge is calculated by assuming that each bond is covalent so that one electron is assigned to each bonded atom. For example, in ammonium ion (NH_4^+) the oxidation state of nitrogen is −3, as all eight valence electrons are assigned to the nitrogen atom that is more electronegative than hydrogen. However, the formal charge is +1, calculated by assigning only four valence electrons (one per bond) to nitrogen. For comparison, the nitrogen in ammonia (NH_3) has oxidation state −3 also but a formal charge of zero. On protonation of ammonia to form ammonium, the formal charge on nitrogen changes, but its oxidation state does not.

Oxidation Number in Naming of Inorganic Compounds

In the nomenclature of inorganic compounds, the oxidation number is represented by a Roman numeral. The oxidation number is equal to the oxidation state using the rules, although they acknowledge other methods can be used. Oxidation numbers must be positive or negative integers, fractional oxidation numbers should not be used, and in the event of any uncertainty alternative naming conventions should be used.

Use in Nomenclature

In older literature the term is referred to as Stock number. However, the use of this term is no longer recommended by IUPAC. The oxidation state in compound naming is placed either as a right superscript to the element symbol in chemical formula, for example Fe^{III}, or in parentheses after the name of the element, iron(III) in chemical names. For example, $Fe_2(SO_4)_3$ is named iron(III) sulfate and its formula can be shown as $Fe_2^{III}(SO_4)_3$. This is because a sulfate ion has a charge of −2, so each iron atom takes a charge of +3. Note that fractional oxidation numbers should not be used in naming. Magnetite, Fe_3O_4, is represented as iron(II,III) oxide, showing the mixture of oxidation states of the nonequivalent iron atoms.

Oxidation Number in Coordination Compounds

While *oxidation state* and *oxidation number* are often used interchangeably, *oxidation number* is used in coordination chemistry with a slightly different meaning. In coordination chemistry, the rules used for counting electrons are different. Every electron in a metal–ligand bond belongs to the ligand, regardless of electronegativity, so that the oxidation number is the charge that would remain if all ligands were removed together with the electron pairs shared with the central atom.

The current IUPAC definition of the oxidation number in a coordination compound is as follows:

Of a central atom in a coordination entity, the charge it would bear if all the ligands were removed along with the electron pairs that were shared with the central atom. It is represented by a Roman numeral.

For most coordination complexes, the metal atom is the less electronegative end of each metal–ligand bond, so that this rule gives the same result as the electronegativity-based rule There are exceptions, however, such as Wilkinson's catalyst, $RhCl(PPh_3)_3$ (Ph = phenyl), in which the rhodium atom is more electronegative than phosphorus. Nevertheless, the oxidation *number* of rhodium in this molecule is considered to be +1 and the molecule's systematic name is chlorotris(triphenylphosphine)rhodium(I), as the electrons of each Rh–P bond are assigned to the P atom of the ligand. The electronegativity rule would assign them instead to the Rh with an oxidation *state* of −5.

Spectroscopic Oxidation States Versus Oxidation Numbers

Although oxidation numbers can be helpful for classifying compounds, they are unmeasurable and their physical meaning can be ambiguous. Oxidation numbers require particular caution for molecules where the bonding is covalent, since the oxidation numbers require the heterolytic removal of ligands, which in essence denies covalency. *Spectroscopic oxidation states*, as defined by Jorgenson and reiterated by Wieghardt, are measurables that are benchmarked using spectroscopic and crystallographic data.

Oxidation state can also have effect on spectroscopic studies of compounds. In infrared spectroscopy of metal carbonyls this effect is illustrated by using spectroscopic studies on metals from oxidation states of −2 to +2.

Unusual Oxidation States

Unusual oxidation states of metals are important in biochemical processes, the notable ones being Fe(IV) and Fe(V) in cytochrome P450-containing systems.

History of the Oxidation Number Concept

First Study

Oxidation itself was first studied by Antoine Lavoisier, who believed that what we now call oxidation was always the result of reactions with oxygen, thus the name. Although Lavoisier's idea has been shown to be incorrect, the name he proposed is still used, albeit more generally.

Oxidation states were one of the intellectual stepping stones that Mendeleev used to derive the periodic table.

Nomenclature

When it was realized that some metals form two different binary compounds with the same non-metal, the two compounds were often distinguished by using the ending *-ic* for the higher metal oxidation state and the ending *-ous* for the lower. For example, $FeCl_3$ is ferric chloride and $FeCl_2$ is ferrous chloride. This system is not very satisfactory (although sometimes still used) because different metals have different oxidation states which have to be learned: ferric and ferrous are +3 and +2 respectively, but cupric and cuprous are +2 and +1, and stannic and stannous are +4 and +2. Also there was no allowance for metals with more than two oxidation states, such as vanadium with oxidation states +2, +3, +4 and +5.

This system has been largely replaced by Stock nomenclature (named for Alfred Stock who suggested it in 1919). Under the Stock system $FeCl_2$ is called iron(II) chloride rather than ferrous chloride.

Current Concept

The current concept of "oxidation state" was introduced by W. M. Latimer in 1938. In 1940 IUPAC recommended that the term Stock number should be replaced by the term oxidation number. In 1947 Linus Pauling proposed that the oxidation number could be determined using the electronegativity of the atoms to determine the "ions" in the formal determination of oxidation number. In 1970 IUPAC defined oxidation number in terms of electronegativity. In 1990 IUPAC changed course and adopted a rule based determination for the "central atom" rather than using electronegativity. This is the definition in the current Gold Book for "oxidation state". They also introduced the definition of oxidation number, shown in the current Gold Book, that appears to make oxidation number specific to coordination chemistry. This may not have been their intention, as in 2005 they issued new recommendations for inorganic nomenclature that define oxidation number in the same terms as the 1990 definition of oxidation state, and that oxidation number is, as in the earlier recommendations, used in the naming of inorganic compounds.

Oxidation Number Versus Oxidation State

In general, in the wider field of chemistry the IUPAC definitions have not been adhered to and both terms are used interchangeably, as they were when Latimer introduced the concept in 1938. For example, two well-known textbooks use the term oxidation state and represent it in Roman numerals in chemical formulae. The point has been made that, if there is any semantic difference between the terms, then oxidation number refers to the specific numerical value assigned to the entity known as oxidation state, much as IUPAC use the term charge number to refer to the numerical value assigned to the entity known as ionic charge. The IUPAC Gold Book takes the definitions from 1990 IUPAC papers rather than the more recent current IUPAC 2005 recommendations. An IUPAC project, "Towards a comprehensive definition of oxidation state", (project 2008-040-1-200) started in 2009. The technical report has been published in *Pure and Applied Chemistry*. The project was undertaken because the current definition in the IUPAC Gold Book was seen to be "narrow and circular", and "inapplicable to clusters, Zintl phases and some organometallic complexes".

Latimer Diagram

A Latimer diagram of an element is a summary of the standard electrode potential data of that element. This type of diagrams is named after Wendell Mitchell Latimer, an American chemist.

Construction

In a Latimer diagram, the most highly oxidized form of the element is on the left, with successively lower oxidation states to the right (e.g. for oxygen, the species would be in the order O_2 (0), H_2O_2 (-1), H_2O (-2)). The species are connected by an arrow, and the numerical value of the standard potential (in volts) for the reduction is written over the arrow.

For example, if the arrow between O_2 and H_2O_2 has a value +0.70 over it, it indicates that the standard electrode potential for the reaction $O_2(g) + 2H^+ + 2e^- \rightleftarrows H_2O_2(aq)$ is +0.70V.

Application

Latimer diagrams can be used in the construction of Frost diagrams, as a concise summary of the standard electrode potentials relative to the element. Since $\Delta_r G° = -vFE°$, the electrode potential is a representation of the Gibbs energy change for the given reduction. The sum of the Gibbs energy changes for subsequent reductions (e.g. from O_2 to H_2O_2, then from H_2O_2 to H_2O) is the same as the Gibbs energy change for the overall reduction (i.e. from O_2 to H_2O), in accordance with Hess's law. This can be used to find the electrode potential for non-adjacent steps, which gives all the information necessary for the Frost diagram.

A simple examination of a Latimer diagram can also indicate if a species will disproportionate in solution under the conditions for which the electrode potentials are given: if the potential to the right of the species is higher than the potential on the left, it will disproportionate.

Utility of the Diagram

The standard reduction potential for the reduction half-reaction involving the two species joined by the arrow is shown above the arrow. Latimer diagrams show the redox information about a series of species in a very condensed form. From these diagrams you can predict the redox behaviour of a given species. The more positive the standard reduction potential, the more readily the species on the left is reduced to the species on the right side of the arrow. Thus, highly positive standard reduction potentials indicate that the species at the left is a good oxidizing agent. Negative standard reduction potentials indicate that the species to the right behaves as a reducing agent.

Construction: In a Latimer diagram of an element, oxidation numbers of the element decrease from left to right and the numerical values of E° in volts) are written above the line joining the species involved in the couple.

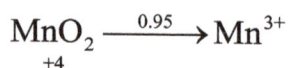

$$MnO_2 \xrightarrow{0.95} Mn^{3+}$$
$$_{+4}$$

In the above example, the notation denotes the half-cell reaction with E° = +0.95 volt

$$MnO_2 + 4H^+ + e \longrightarrow Mn^{3+} + 2H_2O$$

Note that both of these half-reactions involve hydrogen ions, and therefore the potentials depend on pH.

e.g. The Latimer diagram for chlorine in acidic solution, for instance, is

$$ClO_4^- \xrightarrow{+1.2} ClO_3^- \xrightarrow{+1.18} HClO_2 \xrightarrow{+1.65} HClO \xrightarrow{+1.67} Cl_2 \xrightarrow{+1.36} Cl^-$$
$$\;\;\;\;+7 \qquad\qquad +5 \qquad\qquad +3 \qquad\qquad +1 \qquad\qquad 0 \qquad\qquad -1$$

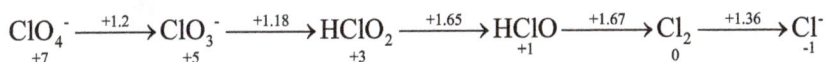

In basic aqueous solution, the Latimer diagram for chlorine is

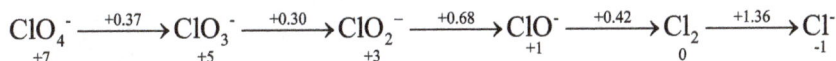

$$ClO_4^- \xrightarrow{+0.37} ClO_3^- \xrightarrow{+0.30} ClO_2^- \xrightarrow{+0.68} ClO^- \xrightarrow{+0.42} Cl_2 \xrightarrow{+1.36} Cl^-$$
$$\;\;\;\;+7 \qquad\qquad +5 \qquad\qquad +3 \qquad\qquad +1 \qquad\qquad 0 \qquad\qquad -1$$

Nonadjacent Species

For the nonadjacent species the standard potential of a couple that is the combination of two other couples is obtained by combining the standard Gibbs energies, not the standard potentials, of the half-reactions.

$$\Delta G^0(a+b) = \Delta G^0(a) + \Delta G^0(b)$$

And as G = -nFE°, hence

$$E^\circ(a+b)$$
$$= \frac{n(a)E^\circ(a) + n(b)E^\circ(b)}{n(a) + n(b)}$$

Disproportionation

Disproportionation is a specific type of redox reaction in which an element from a reaction undergoes both oxidation and reduction to form two different products.

Overview

For example, the UV photolysis of mercury(I) chloride $Hg_2Cl_2 \rightarrow Hg + HgCl_2$ is a disproportionation. Mercury(I) is a diatomic dication Hg_2^{2+}. In this reaction, the chemical bond in the molecular ion is broken, and one mercury atom is reduced to mercury(0), and the other is oxidized to mercury(II).

A similar type of reaction, but in which no element changes oxidation number, is the acid-base disproportionation reaction observed when an amphiprotic species reacts with itself. Two common examples for conjugated bases of polyprotic acids such as bicarbonate and dihydrogenophosphate are respectively:

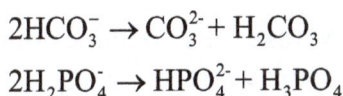

$$2HCO_3^- \rightarrow CO_3^{2-} + H_2CO_3$$
$$2H_2PO_4^- \rightarrow HPO_4^{2-} + H_3PO_4$$

The oxidation numbers remain constant in these acid-base reactions: O = −2, H = +1, C = +4, P = +5. This is also called autoionization.

Another variant on disproportionation is radical disproportionation, in which two radicals form an alkane and alkene.

Reverse Reaction

The reverse of disproportionation, when a compound in an intermediate oxidation state is formed from compounds in lower and higher oxidation states, is called comproportionation, also known as symproportionation.

History

The first disproportionation reaction to be studied in detail was:

$$2\,Sn^{2+} \rightarrow Sn^{4+} + Sn$$

This was examined using tartrates by Johan Gadolin in 1788. In the Swedish version of his paper he called it 'söndring'.

Examples

- Chlorine gas reacts with dilute sodium hydroxide to form sodium chloride, sodium chlorate and water. The ionic equation for this reaction is as follows:

$$3\,Cl_2 + 6\,OH^- \rightarrow 5\,Cl^- + ClO_3^- + 3\,H_2O$$

  - The chlorine gas reactant is in oxidation state 0. In the products, the chlorine in the Cl^- ion has an oxidation number of -1, having been reduced, whereas the oxidation number of the chlorine in the ClO_3^- ion is $+5$, indicating that it has been oxidized.

- Bromine fluoride undergoes disproportionation reaction for form bromine trifluoride:

$$3\,BrF \rightarrow BrF_3 + Br_2$$

- The dismutation of superoxide free radical to hydrogen peroxide and oxygen, catalysed in living systems by the enzyme superoxide dismutase:

$$2\,O_2^- + 2\,H^+ \rightarrow H_2O_2 + O_2$$

  The oxidation state of oxygen is $-1/2$ in the superoxide free radical anion, -1 in hydrogen peroxide and 0 in dioxygen.

- In the Cannizzaro reaction, an aldehyde is converted into an alcohol and a carboxylic acid. In the related Tishchenko reaction, the organic redox reaction product is the corresponding ester. In the Kornblum–DeLaMare rearrangement, a peroxide is converted to a ketone and an alcohol.

- The disproportionation of hydrogen peroxide into water and oxygen catalysed by either potassium iodide or the enzyme catalase:

$$2\,H_2O_2 \rightarrow 2\,H_2O + O_2$$

- The Boudouard reaction is for example used in the HiPco method for producing carbon nanotubes, high-pressure carbon monoxide disproportionates when catalysed on the surface of an iron particle:

$$2\,CO \rightarrow C + CO_2$$

- Nitrogen has oxidation state +IV in nitrogen dioxide, but when this compound reacts with water, it forms both nitric acid and nitrous acid, where nitrogen has oxidation states +V and +III respectively:

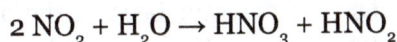

$$2\,NO_2 + H_2O \rightarrow HNO_3 + HNO_2$$

Biochemistry

In 1937, Hans Adolf Krebs, who discovered the citric acid cycle bearing his name, confirmed the anaerobic dismutation of pyruvic acid in lactic acid, acetic acid and CO_2 by certain bacteria according to the global reaction:

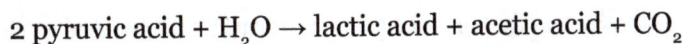

$$2\ \text{pyruvic acid} + H_2O \rightarrow \text{lactic acid} + \text{acetic acid} + CO_2$$

The dismutation of pyruvic acid in other small organic molecules (ethanol + CO_2, or lactate and acetate, depending on the environmental conditions) is also an important step in fermentation reactions. Fermentation reactions can also be considered as disproportionation or dismutation biochemical reactions. Indeed, the donor and acceptor of electrons in the redox reactions supplying the chemical energy in these complex biochemical systems are the same organic molecules simultaneously acting as reductant or oxidant.

Another example of biochemical dismutation reaction is the disproportionation of acetaldehyde into ethanol and acetic acid.

While in respiration electrons are transferred from substrate (electron donor) to an electron acceptor, in fermentation part of the substrate molecule itself accepts the electrons. Fermentation is therefore a type of disproportionation, and does not involve an overall change in oxidation state of the substrate. Most of the fermentative substrates are organic molecules. However, a rare type of fermentation may also involve the disproportionation of inorganic sulfur compounds in certain sulfate-reducing bacteria.

Frost Diagram

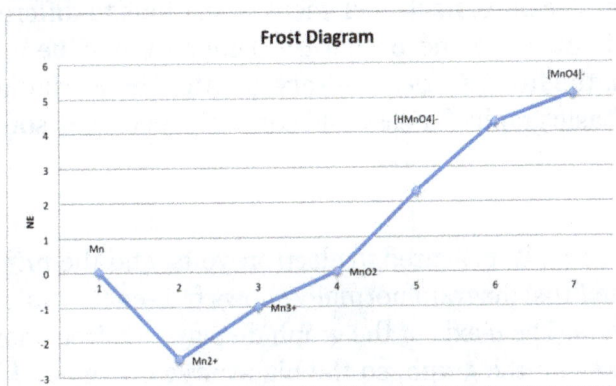

An example of a Frost diagram for the element manganese

A Frost diagram or Frost-Ebsworth diagram is a type of graph used by inorganic chemists in electrochemistry to illustrate the relative stability of a number of different oxidation states of a particular substance. The graph illustrates the oxidation state vs free energy of a chemical species. This effect is dependent on pH, so this parameter also must be included. The free energy is determined by the oxidation–reduction half-reactions. The Frost diagram allows easier comprehension of these reduction potentials than the earlier-designed Latimer diagram, because the "lack of additivity of potentials" was confusing. The free energy $\Delta G°$ is related to reduction potential E in the graph by given formula: $\Delta G° = -nFE°$ or $nE° = -\Delta G°/F$, where n is the number of transferred electrons, and F is Faraday constant ($F = 96,485$ J/(V·mol)). The Frost diagram is named for Arthur Atwater Frost, who originally created them as a way to "show both free energy and oxidation potential data conveniently" in a 1951 paper.

pH Dependence

The pH dependence is given by the factor $-0.059m/n$ per pH unit, where m relates to the number of protons in the equation, and n the number of electrons exchanged. Electrons are always exchanged in electrochemistry, but not necessarily protons. If there is no proton exchange in the reaction equilibrium, the reaction is said to be *pH-independent*. This means that the values for the electrochemical potential rendered in a redox half-reaction, whereby the elements in question change oxidation states are the same whatever the pH conditions under which the procedure is carried out.

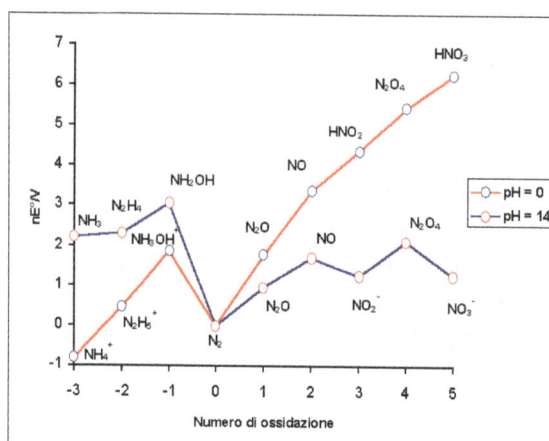

Frost diagram for nitrogen at different pH levels

The Frost diagram is also a useful tool for comparing the trends of standard potentials (slope) of acidic and basic solutions. The pure, neutral element transitions to different compounds depending whether the species is in acidic and basic pHs. Though the value and amount of oxidation states remain unchanged, the free energies can vary greatly. The Frost diagram allows the superimposition of acidic and basic graphs for easy and convenient comparison.

Unit and Scale

The standard free-energy scale is measured in electron-volts, and the $nE° = 0$ value is usually the pure, neutral element. The Frost diagram normally shows free-energy values above and below $nE° = 0$ and is scaled in integers. The y axis of the graph displays the free energy. Increasing stability (lower free energy) is lower on the graph, so the higher free energy and higher on the graph an element is, the more unstable and reactive it is.

The oxidation state of the element is shown on the x axis of the Frost diagram. Oxidation states are unitless and are also scaled in positive and negative integers. Most often, the Frost diagram displays oxidation number in increasing order, but in some cases it is displayed in decreasing order. The neutral, pure element with a free energy of zero ($nE° = 0$) also has an oxidation state equal to zero.

The slope of the line therefore represents the standard potential between two oxidation states. In other words, the steepness of the line shows the tendency for those two reactants to react and form the lowest-energy product. There is a possibility of having either a positive or negative slope. A positive slope between two species indicates a tendency for an oxidation reaction, while a negative slope between two species indicates a tendency for reduction. For example, if the manganese in $[HMnO_4]^-$ has an oxidation state of +6 and $nE° = 4$, and in MnO_2 the oxidation state is +4 and $nE° = 0$, then the slope $\Delta y/\Delta x$ is 4/2 = 2, yielding the standard potential of +2. The stability of any terms can be similarly found by this graph.

Gradient

The gradient of the line between any two points on a Frost diagram gives the potential for the reaction. A species that lies in a peak, above the gradient of the two points on either side, denotes a species unstable with respect to disproportionation, and a point that falls below the gradient of the line joining its two adjacent points lies in a thermodynamic sink, and is *intrinsically stable*.

Axes

The axes of the Frost diagram show (horizontally) the oxidation state of the species in question and (vertically) the electron exchange number multiplied by the voltage (nE) or the Gibbs free energy per unit of the Faraday constant, $\Delta G/F$.

Disproportionation and Comproportionation

In regards to electrochemical reactions, two main types of reactions can be visualized using the Frost diagram. Comproportionation is when two equivalents of an element, differing in oxidation number, combine to form a product with an intermediate oxidation number. Disproportionation is the opposite reaction, in which two equivalents of an element, identical in oxidation number, react to form two products of differing oxidation numbers.

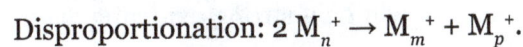

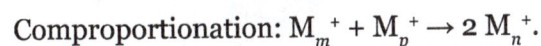

Disproportionation: $2\,M_n^+ \rightarrow M_m^+ + M_p^+$.

Comproportionation: $M_m^+ + M_p^+ \rightarrow 2\,M_n^+$.

$2n = m + p$ in both examples.

Using a Frost diagram, one can predict whether one oxidation number would undergo disproportionation or two oxidation numbers would undergo comproportionation. Looking at two slopes among a set of three oxidation numbers on the diagram, assuming the two standard potentials (slopes) are not equal, the middle oxidation will either be in a "hill" or "valley" form. A hill is formed when the left slope is steeper than the right, and a valley is formed when the right slope is steeper than the left. An oxidation number that is on "top of the hill" tends to favor disproportionation into the adjacent oxidation states. The adjacent oxidation states, however, will favor compro-portionation if the middle oxidation state is in the "bottom of a valley".

Criticisms/discrepancies

Arthur Frost stated in his own original publication that there may be potential criticism for his Frost diagram. He predicts that "the slopes may not be as easily or accurately recognized as they are the direct numerical values of the oxidation potentials [of the Latimer diagram]". Many inorganic chemists use both the Latimer and Frost diagrams in tandem, using the Latimer for quantitative data, and then converting those data into a Frost diagram for visualization. Frost suggested that the numerical values of standard potentials could be added next to the slopes to provide supplemental information.

In a paper by Jesús M. Martinez de Ilarduya, he warns users of Frost diagrams to be aware of the definition of free energy being used to construct the diagrams. In acid-solution graphs, the standard $nE°= -\Delta G/F$ is universally used; therefore all sources' acid-solution Frost diagrams will be identical. However, various textbooks show discrepancies in the Frost diagram of an element, in regards to the energy. Some textbooks use the same reduction potential ($E°(H^+/H_2)$) as an acid-solution for a basic-solution. In the Phillips and Williams Inorganic Chemistry textbook, however, a new reduction potential is used for the basic solutions given by the following formula: $E°(OH) = E°_b - E°(H_2O/H_2OH^-) = E°_b + 0.828$. This new type of reduction potential is used in some textbooks and not others, and is not always notated on the graph. Users of the Frost diagram should be aware of which free-energy scale their diagram displays.

Utility and Limitation of Frost Diagram

- Thermodynamic stability is found at the bottom of the diagram. Thus, the lower a species is positioned on the diagram, the more thermodynamically stable it is (from a oxidation-reduction perspective) Mn (II) is the most stable species.

- A species located on a convex curve can undergo disproportionation MnO_4^{2-} and Mn (III) tends to disproportionate.

- Those species on a concave curve do not typically disproportionate. MnO_2 does not disproportionate

- Any species located on the upper left side of the diagram will be a strong oxidizing agent MnO_4^- is a strong oxidizer.

- Any species located on the upper rignt side of the diagram will be a reducing agent.

- manganese metal is a moderate reducing agent

- These diagrams describe the thermodynamic stability of the various species. Although a given species might be thermodynamically unstable toward reduction, the kinetics of such a reaction might be very slow. Although it is thermodynamically favourable for permanganate ion to be reduced to Mn (II) ion, the reaction is slow except in the presence of a catalyst. Thus, solutions of permanganate can be stored and used in the laboratory

- The information obtained from a Frost diagram is for species under standard conditions (pH=0 for acidic solution and pH=14 for basic solution). Changes in pH may change the relative stabilities of the species. The potential of any process involving the hydrogen ion will change with pH because the concentration of this species is changing. Under basic conditions aqueous Mn^{2+} does not exist. Instead Insoluble $Mn(OH)_2$ forms

Redox Principles Involved in Extraction of Elements

Electrochemical cells:

A device producing an electric current from a chemical reaction (Redox reaction) is called electrochemical cell i.e. it converts chemical energy to electrical energy.

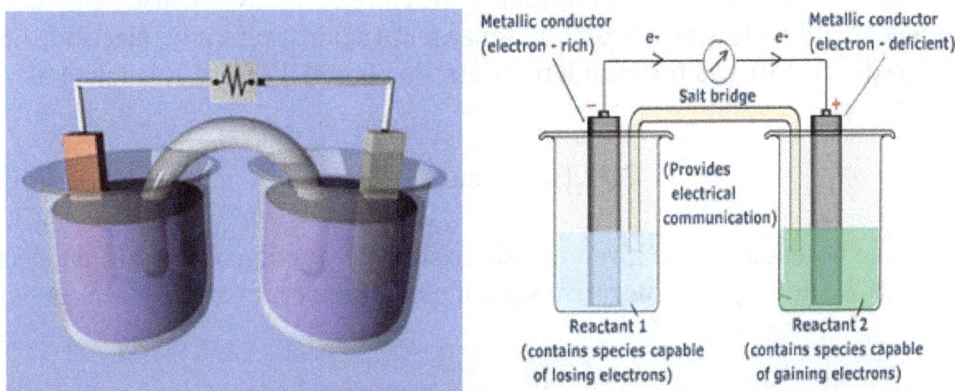

A demonstration electrochemical cell setup resembling the Daniell cell. The two half-cells are linked by a salt bridge carrying ions between them. Electrons flow in the external circuit.

A redox reaction consists of two half reactions called oxidation half reaction and reduction half reaction.

$$Zn + CuSO_4 \longrightarrow ZnSO_4 + Cu \text{ or } Zn + Cu^{2+} \longrightarrow Zn^{2+} + Cu$$

Two half redox reactions are

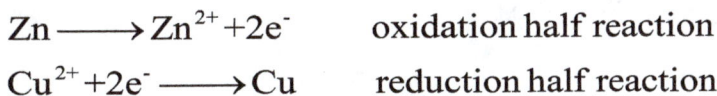

$$Zn \longrightarrow Zn^{2+} + 2e^- \qquad \text{oxidation half reaction}$$
$$Cu^{2+} + 2e^- \longrightarrow Cu \qquad \text{reduction half reaction}$$

Electrochemical cell based on this reaction is called Daniel cell.

In electrochemical cell those two half reactions are used as two half cells joined together by salt bridge.

Salt bridge: A U-shaped tube containing a concentrated solution of an inert electrolyte like K_2SO_4, KCl, KNO_3 etc. The function of salt bridge is to complete inert circuit by flow of ions and to maintain the electrical neutrality in the solution of two half cells.

For making the salt bridge only those electrolytes are used for which cations and anions have nearly same ionic mobility.

In electrochemical cell, the electrode at which oxidation takes place is called anode or negative pole and the electrode at which reduction takes place is called anode or positive pole.

Electrochemical cell is represented as electron flow from anode to cathode in the external circuit while current flows from cathode to anode.

Electrode potential or half-cell potential:

The tendency of an electrode to lose electrons is called oxidation potential, while tendency of an electrode to gain electrons is called reduction potential. Electrode potential depends upon concentration of metal ion and temperature. At standard condition i.e. 1 molar concentration and 298k are called standard electrode potential.

The absolute value of electrode potential cannot be measured directly because half reaction cannot take place independently. Electrode potential is measured against reference electrode i.e. standard hydrogen electrode (S.H.E) and the standard electrode potential of S.H.E is taken as zero. S.H.E is represented as

$$Pt, H_2(gas, 1\ atm) \,|\, H^+(1M)$$

Standard electrode potential (E°) is given a positive sign if reduction occurs at that electrode with respect to the S.H.E and is given a negative sign, if oxidation occurs at the electrode with respect to the S.H.E.

EMF of Cell and Electrode Potential

The electromotive force or EMF of a cell is expressed in term of potential difference established between the half cells when no current passes through the cell. Under standard condition of temperature and concentration EMF is known as standard EMF and abbreviated as E° . This is expressed in volts.

$$E° = [standard\ reduction\ potental\ of\ cathode] - [standard\ oxidation\ potential\ of\ anode]$$
$$= E°_{cathode} - E°_{anode} = E°_{right} - E°_{left}$$

Nernst Equation

In electrochemistry, the Nernst equation is an equation that relates the reduction potential of an electrochemical reaction (half-cell or full cell reaction) to the standard electrode potential, temperature, and activities (often approximated by concentrations) of the chemical species undergoing reduction and oxidation. It is the most important equation in the field of electrochemistry. It is named after the German physical chemist who first formulated it, Walther Nernst.

Expression

The Nernst equation is easily derived from the standard changes in the Gibbs free energy associated with an electrochemical transformation. For any electrochemical reduction reaction of the form $Ox + ze \rightarrow Red$, standard thermodynamics says that the actual free energy change ΔG is related to the free energy change under standard conditions $\Delta G^{\ominus}$ by the relationship $\Delta G = \Delta G^{\ominus} + RT \ln(Q)$, where Q is the reaction quotient. The electrochemical potential E associated with an electrochemical reaction of the form $Ox + ze \rightarrow Red$ is defined as the decrease in Gibbs free energy per Coulomb of charge transferred, which leads to the relationship $\Delta G = -zFE$. The constant F (the Faraday constant) is a units conversion factor $F = N_A q$ where NA is Avogadro's number and q is the fundamental electron charge. This immediately leads to the Nernst Equation.

The Nernst equation for an electrochemical half-cell is:

$$E_{red} = E_{red}^{\ominus} - \frac{RT}{zF} \ln \mathbb{Q} = E_{red}^{\ominus} - \frac{RT}{zF} \ln \frac{a_{Red}}{a_{Ox}}$$

For a complete electrochemical reaction (full cell), the equation can also be written as:

$$E_{cell} = E_{cell}^{\ominus} - \frac{RT}{zF} \ln Q_r \text{ (total cell potential)}$$

where

- E_{red} is the half-cell reduction potential at the temperature of interest

- E⊖ red is the *standard* half-cell reduction potential

- E_{cell} is the cell potential (electromotive force) at the temperature of interest

- E⊖ cell is the *standard* cell potential

- R is the universal gas constant: $R = 8.314472(15)$ J K^{-1} mol^{-1}

- T is the temperature in kelvins

- a is the chemical activity for the relevant species, where a_{Red} is the activity of the reduced form and a_{Ox} is the activity of the oxidized form. Similarly to equilibrium constants, activities are always measured with respect to the standard state (1M for solutes, 1 atm for gases). The activity of species x, a_x, can be related to the physical concentrations c_x via $a_x = \gamma_x c_x$, where γ_x is the activity coefficient of species X. Because activity coefficients tend to unity at low concentrations, activities in the Nernst equation are frequently replaced by simple concentrations.

- F is the Faraday constant, the number of coulombs per mole of electrons: $F = 9.64853399(24) \times 10^4$ C mol^{-1}

- z is the number of moles of electrons transferred in the cell reaction or half-reaction

- Q_r is the reaction quotient.

At room temperature $(25°C)$, $\dfrac{RT}{F}$ may be treated like a constant and replaced by 25.693 mV for cells.

The Nernst equation is frequently expressed in terms of base 10 logarithms (*i.e.*, common logarithms) rather than natural logarithms, in which case it is written, *for a cell at 25 °C*:

$$E = E^0 + \frac{0.05916}{z} \log_{10} \frac{a_{\text{Ox}}}{a_{\text{Red}}}.$$

The Nernst equation is used in physiology for finding the electric potential of a cell membrane with respect to one type of ion.

Nernst Potential

The Nernst equation has a physiological application when used to calculate the potential of an ion of charge z across a membrane. This potential is determined using the concentration of the ion both inside and outside the cell:

$$E = \frac{RT}{zF} \ln \frac{[\text{ion outside cell}]}{[\text{ion inside cell}]} = 2.3026 \frac{RT}{zF} \log_{10} \frac{[\text{ion outside cell}]}{[\text{ion inside cell}]}.$$

When the membrane is in thermodynamic equilibrium (i.e., no net flux of ions), the membrane potential must be equal to the Nernst potential. However, in physiology, due to active ion pumps, the inside and outside of a cell are not in equilibrium. In this case, the resting potential can be determined from the Goldman equation:

$$E_m = \frac{RT}{F} \ln \left(\frac{\sum_i^N P_{M_i^+}[M_i^+]_{\text{out}} + \sum_i^M P_{A_j^-}[A_j^-]_{\text{in}}}{\sum_i^N P_{M_i^+}[M_i^+]_{\text{in}} + \sum_i^M P_{A_j^-}[A_j^-]_{\text{out}}} \right)$$

- E_m is the membrane potential (in volts, equivalent to joules per coulomb)
- P_{ion} is the permeability for that ion (in meters per second)
- $[\text{ion}]_{\text{out}}$ is the extracellular concentration of that ion (in moles per cubic meter, to match the other SI units, though the units strictly don't matter, as the ion concentration terms become a dimensionless ratio)
- $[\text{ion}]_{\text{in}}$ is the intracellular concentration of that ion (in moles per cubic meter)
- R is the ideal gas constant (joules per kelvin per mole)
- T is the temperature in kelvins
- F is Faraday's constant (coulombs per mole)

The potential across the cell membrane that exactly opposes net diffusion of a particular ion through the membrane is called the Nernst potential for that ion. As seen above, the magnitude of the Nernst potential is determined by the ratio of the concentrations of that specific ion on the two sides of the membrane. The greater this ratio the greater the tendency for the ion to diffuse in one direction, and therefore the greater the Nernst potential required to prevent the diffusion.

A similar expression exists that includes r (the absolute value of the transport ratio). This takes transporters with unequal exchanges into account. Sodium-potassium pump where the transport ratio would be $\frac{2}{3}$. The other variables are the same as above. The following example includes two ions: potassium (K^+) and sodium (Na^+). Chloride is assumed to be in equilibrium.

$$V_m = \frac{RT}{F} \ln \left(\frac{rP_{K^+}[K^+]_{out} + P_{Na^+}[Na^+]_{out}}{rP_{K^+}[K^+]_{in} + P_{Na^+}[Na^+]_{in}} \right)$$

When chloride (Cl^-) is taken into account, its part is flipped to account for the negative charge.

$$V_m = \frac{RT}{F} \ln \left(\frac{P_{K^+}[K^+]_{out} + P_{Na^+}[Na^+]_{out} + P_{Cl^-}[Cl^-]_{in}}{P_{K^+}[K^+]_{in} + P_{Na^+}[Na^+]_{in} + P_{Cl^-}[Cl^-]_{out}} \right)$$

Derivation

Using Boltzmann Factors

For simplicity, we will consider a solution of redox-active molecules that undergo a one-electron reversible reaction

$$Ox + e^- \rightleftharpoons Red$$

and that have a standard potential of zero. The chemical potential μ_c of this solution is the difference between the energy barriers for taking electrons from and for giving electrons to the working electrode that is setting the solution's electrochemical potential.

The ratio of oxidized to reduced molecules, $\frac{[Ox]}{[Red]}$, is equivalent to the probability of being oxidized (giving electrons) over the probability of being reduced (taking electrons), which we can write in terms of the Boltzmann factor for these processes:

$$\frac{[Red]}{[Ox]} = \frac{\exp(-[\text{barrier for gaining an electron}]/kT)}{\exp(-[\text{barrier for losing an electron}]/kT)} = \exp\left(\frac{\mu_c}{kT}\right).$$

Taking the natural logarithm of both sides gives

$$\mu_c = kT \ln \frac{[Red]}{[Ox]}.$$

If $\mu_c \neq 0$ at $\frac{[Red]}{[Ox]} = 1$, we need to add in this additional constant:

$$\mu_c = \mu_c^0 + kT \ln \frac{[Red]}{[Ox]}.$$

Dividing the equation by e to convert from chemical potentials to electrode potentials, and remembering that $\frac{k}{e} = \frac{R}{F}$ we obtain the Nernst equation for the one-electron process $Ox + e^- \rightarrow Red$:

$$E = E^0 + \frac{kT}{e} \ln \frac{[\text{Red}]}{[\text{Ox}]}$$

$$= E^0 - \frac{RT}{F} \ln \frac{[\text{Ox}]}{[\text{Red}]}.$$

Using Thermodynamics (Chemical Potential)

Quantities here are given per molecule, not per mole, and so Boltzmann constant k and the electron charge e are used instead of the gas constant R and Faraday's constant F. To convert to the molar quantities given in most chemistry textbooks, it is simply necessary to multiply by Avogadro's number: $R = kN_A$ and $F = eN_A$.

The entropy of a molecule is defined as

$$S \stackrel{\text{def}}{=} k \ln \Omega,$$

where Ω is the number of states available to the molecule. The number of states must vary linearly with the volume V of the system (here an idealized system is considered for better understanding, so that activities are posited very close to the true concentrations. Fundamental statistical proof of the mentioned linearity goes beyond the scope of this section, but to see this is true it is simpler to consider usual isothermal process for an ideal gas where the change of entropy $\Delta S = nR \ln\left(\frac{V_2}{V_1}\right)$ takes place. It follows from the definition of entropy and from the condition of constant temperature and quantity of gas n that the change in the number of states must be proportional to the relative change in volume $\frac{V_2}{V_1}$. In this sense there is no difference in statistical properties of ideal gas atoms compared with the dissolved species of a solution with activity coefficients equaling one: particles freely "hang around" filling the provided volume), which is inversely proportional to the concentration c, so we can also write the entropy as

$$S = k \ln (\text{constant} \times V) = -k \ln (\text{constant} \times c).$$

The change in entropy from some state 1 to another state 2 is therefore

$$\Delta S = S_2 - S_1 = -k \ln \frac{c_2}{c_1},$$

so that the entropy of state 2 is

$$S_2 = S_1 - k \ln \frac{c_2}{c_1}.$$

If state 1 is at standard conditions, in which c_1 is unity (e.g., 1 atm or 1 M), it will merely cancel the units of c_2. We can, therefore, write the entropy of an arbitrary molecule A as

$$S(A) = S^0(A) - k\ln[A],$$

where S^0 is the entropy at standard conditions and [A] denotes the concentration of A.

The change in entropy for a reaction

$$a\,A + b\,B \rightarrow y\,Y + z\,Z$$

is then given by

$$\Delta S_{rxn} = \big(yS(Y) + zS(Z)\big) - \big(aS(A) + bS(B)\big) = \Delta S_{rxn}^0 - k\ln\frac{[Y]^y[Z]^z}{[A]^a[B]^b}.$$

We define the ratio in the last term as the reaction quotient:

$$Q_r = \frac{\prod_j a_j^{v_j}}{\prod_j a_i^{v_i}} \approx \frac{[Z]^z[Y]^y}{[A]^a[B]^b}.$$

where the numerator is a product of reaction product activities, a_j, each raised to the power of a stoichiometric coefficient, v_j, and the denominator is a similar product of reactant activities. All activities refer to a time t. Under certain circumstances each activity term such as $a_j^{v_j}$ may be replaced by a concentration term, [A].

In an electrochemical cell, the cell potential E is the chemical potential available from redox reactions ($E = \frac{\mu_c}{e}$). E is related to the Gibbs energy change ΔG only by a constant: $\Delta G = -nFE$, where n is the number of electrons transferred and F is the Faraday constant. There is a negative sign because a spontaneous reaction has a negative free energy ΔG and a positive potential E. The Gibbs energy is related to the entropy by $G = H - TS$, where H is the enthalpy and T is the temperature of the system. Using these relations, we can now write the change in Gibbs energy,

$$\Delta G = \Delta H - T\Delta S = \Delta G^0 + kT\ln Q_r,$$

and the cell potential,

$$E = E^0 - \frac{kT}{ne}\ln Q_r.$$

This is the more general form of the Nernst equation. For the redox reaction $Ox + n\,e^- \rightarrow Red$,

$$Q_r = \frac{[Red]}{[Ox]},$$

and we have:

$$E = E^0 - \frac{kT}{ne} \ln \frac{[\text{Red}]}{[\text{Ox}]}$$

$$= E^0 - \frac{RT}{nF} \ln \frac{[\text{Red}]}{[\text{Ox}]}$$

$$= E^0 - \frac{RT}{nF} \ln Q_r.$$

The cell potential at standard conditions E^0 is often replaced by the formal potential $E^{0\prime}$, which includes some small corrections to the logarithm and is the potential that is actually measured in an electrochemical cell.

Relation to Equilibrium

At equilibrium, the electrochemical potential $E = 0$ and therefore the reaction quotient attains the special value known as the equilibrium constant: $Q = K$. Therefore,

$$0 = E^0 - \frac{RT}{nF} \ln K$$

$$\ln K = \frac{nFE^0}{RT}$$

Or at standard temperature,

$$\log_{10} K = \frac{nE^0}{0.05916 \text{ V}} \quad \text{at } T = 298.15 \text{ K.}$$

We have thus related the standard electrode potential and the equilibrium constant of a redox reaction.

Limitations

In dilute solutions, the Nernst equation can be expressed directly in the terms of concentrations (since activity coefficients are close to unity). But at higher concentrations, the true activities of the ions must be used. This complicates the use of the Nernst equation, since estimation of non-ideal activities of ions generally requires experimental measurements.

The Nernst equation also only applies when there is no net current flow through the electrode. The activity of ions at the electrode surface changes when there is current flow, and there are additional overpotential and resistive loss terms which contribute to the measured potential.

At very low concentrations of the potential-determining ions, the potential predicted by Nernst equation approaches toward $\pm\infty$. This is physically meaningless because, under such conditions, the exchange current density becomes very low, and there is no thermodynamic equilibrium necessary for Nernst equation to hold. The electrode is called to be unpoised in such case. Other effects tend to take control of the electrochemical behavior of the system.

Time Dependence of the Potential

The expression of time dependence has been established by Karaoglanoff.

Significance to Related Scientific Domains

The equation has been involved in the scientific controversy involving cold fusion. The discoverers of cold fusion, Fleischmann and Pons, calculated that a palladium cathode immersed in a heavy water electrolysis cell could achieve up to 10^{27} atmospheres of pressure on the surface of the cathode, enough pressure to cause spontaneous nuclear fusion. In reality, only 10,000–20,000 atmospheres were achieved. John R. Huizenga claimed their original calculation was affected by a misinterpretation of Nernst equation. He cited a paper about Pd–Zr alloys. The equation permits the extent of reaction between two redox systems to be calculated and can be used, for example, to decide whether a particular reaction will go to completion or not.

At equilibrium the emfs of the two half cells are equal. This enables Kc to be calculated hence the extent of the reaction.

Some applications:

Purification of water: Ozone is a strong oxidizing agent and it is capable of destroying organic pollutants and bacteria present in water. Hence, oxidation reaction can be utilize for water purification process.

Electroplating: In this process electrical current is employed to reduce dissolved metal cation followed by deposition on a metal plate. This techniques is high in use to prevent corrosion in metals and its matalic machines.

Metallurgy: Metal ores are obtained in their complex form mainly as oxides, sulfides. Reduction of those ores provides pure metal, e.g. Hematite (Fe_2O_3), magnetite (Fe_3O_4) are reduced in the presence of carbon to have metallic iron.

Redox Dependence

Concentration dependence: The concentration of the reacting species in a half-cell greatly influences the potential of a system. A half-cell reaction is generally expressed as

Oxidized form + ne$^-$ = Reduced form

For this reaction, the Nernst reaction is

$$E = E^0 - \frac{RT}{nF} \ln \frac{[Reduced]}{[Oxidized\ form]}$$

This shows that a tenfold increase in the concentration of the oxidized form will raise the half-cell potential by 0.059/n volts at 298°K.

pH dependence: Many redox reactions in aqueous solution involve transfer of proton as well as electrons and the electrode potential or Value of E° therefore depends on the pH.

When hydrogen or hydroxyl ion or electron are involved in a reaction(half-cell reaction), there concentration appear in the Nertnst equation & accordingly the potential is influenced by the pH of the medium.

A reaction of this type should be represented through this way;

$Ox + ne^- + mH^+ = Red$

For this reaction, $E = E° - (RT/nF)\ln Q$

Where n = numbers of electron transfer

$$Q = \frac{[Red]}{[Ox][H^+]^m}$$

Ultimately at 25° C, equation (i) can be expressed as

$$E = E^0 - \frac{RT}{nF}\ln\frac{\lfloor Red \rfloor}{Ox} + \frac{mRT}{nF}\ln[H^+]$$

As we know that $\ln[H^+]=2.303\log[H^+]$ and $pH=-\log[H^+]$, hence

$$E=E' - \frac{mRT}{nF}pH, \qquad \text{where } E'=E° - \frac{RT}{nF}\ln\frac{\lfloor Red \rfloor}{[Ox]}$$

Q. Write down the simplest form of equation for the given two half-cell reaction?

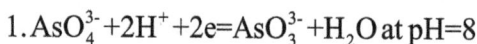

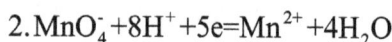

1. $AsO_4^{3-} + 2H^+ + 2e = AsO_3^{3-} + H_2O$ at pH=8

2. $MnO_4^- + 8H^+ + 5e = Mn^{2+} + 4H_2O$

Given that standard reduction potential value E° for AsO_4^{2-}/AsO_3^{2-} & MnO_4^-/Mn^{2+} are +0.56 and 1.51 volt, respectively.

$$\text{Answer:1. } E=+0.088+\frac{0.059}{2}\ln\frac{[AsO_4^{3-}]}{[AsO_3^{3-}]}$$

$$\text{2. } E=+1.51-0.094pH+\frac{0.059}{5}\ln\frac{[MnO_4^{3-}]}{[Mn^{2+}]}$$

Effect of complexation: The formation of a more thermodynamically stable complex when the metal is in the higher oxidation state of a couple favours oxidation and makes the standard potential more negative; the formation of a more stable complex when the metal is in the lower oxidation state of the couple favours reduction and the standard potential becomes more positive.

The formation of metal complexes affects standard potentials because the ability of a complex (ML) formed by coordination of a ligand (L) to accept or release an electron differs from that of the corresponding aqua ion (M).

$$M^{n+}(aq) + e^- \rightarrow M^{(n-1)+}(aq) \qquad E^\circ(M)$$

$$ML^{n+}(aq) + e^- \rightarrow ML^{(n-1)+}(aq) \qquad E^\circ(ML)$$

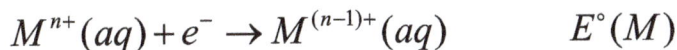

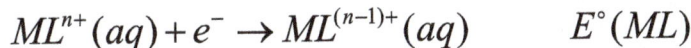

The change in standard potential for the ML redox couple relative to that of M reflects the degree to which the ligand L coordinates more strongly to the oxidized or reduced form of M. The change in standard potential is analysed by considering the thermodynamic cycle shown in Fig. Because the sum of reaction Gibbs energies round the cycle is zero, we can write

$$-FE^\circ(M) - RT \ln K^{ox} + FE^\circ(ML) + RT \ln K^{red} = 0$$

$$E^\circ(M) - E^\circ(ML) = \frac{RT}{F} \ln \frac{K^{ox}}{K^{red}}$$

Where K^{ox} and K^{red} are equilibrium constants for L binding to M^{n+} and $M^{(n-1)+}$ respectively (of the form K = [ML] / [M][L]), and we have used $\Delta rG^\circ = RT \ln K$ in each case.

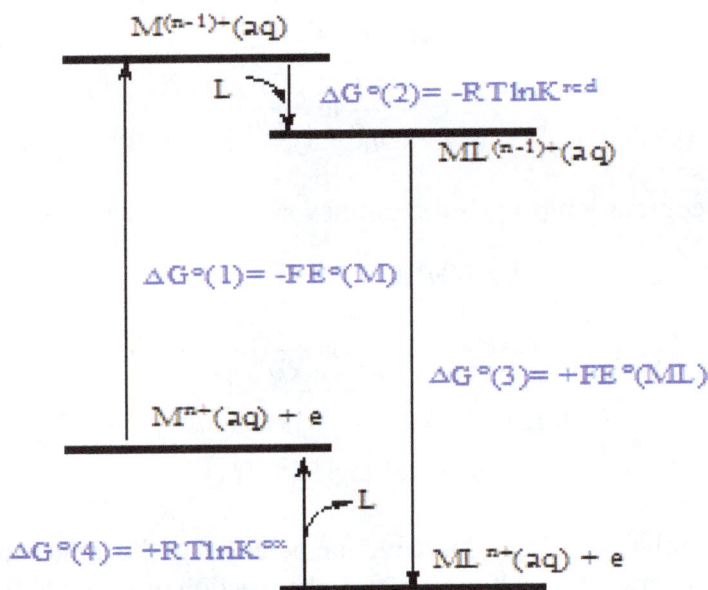

Figure Thermodynamic cycle showing how the standard potential of the couple M⁺/M is altered by the presence of a ligand L.

We know that at 25°,

$$E^\circ(M) - E^\circ(ML) = (0.059V) \ln \frac{K^{ox}}{K^{red}}$$

Thus, every ten-fold increase in the equilibrium constant for ligand binding to M^{n+} compared to $M^{(n-1)+}$ decreases the reduction potential by 0.059 V.

Effect of Precipitation:

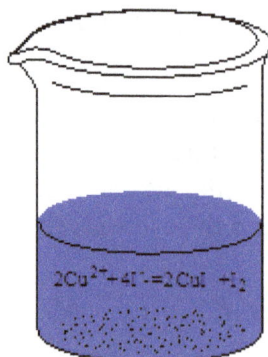

Precipitation is due to change of concentration of the oxidized and reduced form and precipitation of one product is happened due to low solubility i.e. it depend upon the solubility product of that compound, e.g. due to low reduction potential of Cu^+/Cu^{2+} (+0.15 volt), it is not expected to oxidize iodide to iodine ($E° = +0.54$ volt) but in practice due to low solubility of CuI, the concentration of Cu^+ ion reduces in such a way that the reduction potential is substantially increased.

$$Cu^{2+} + e = Cu^+$$

$$E = E° - \frac{RT}{nF} \ln \frac{\left[Cu^+\right]}{\left[Cu^{2+}\right]}$$

The concentration of cuprous ion (Cu^+) is determined by the solubility product (k) of CuI,

$$k = [Cu^+][I^-] = 10^{-12}$$

$$So,\ E = E° - \frac{RT}{nF} \ln \frac{10^{-12}}{\left[Cu^{2+}\right]\left[I^-\right]}$$

$$E = 0.15 + 12 \times 0.059 + 0.059 \log[Cu^{2+}][I^-]$$

$$= +0.86 + 0.059 \log[Cu^{2+}][I^-]$$

The effective potential thus rises above the reduction potential of the iodide-iodine system and iodide ion is oxidized by cupric ion in solution. In fact, the reaction proceeds almost toward the right.

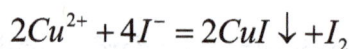

$$2Cu^{2+} + 4I^- = 2CuI \downarrow + I_2$$

Cupper may be estimated by treating the liberated iodine with a solution of sodium thiosulfate.

References

- Loock, Hans-Peter (2011). "Expanded Definition of the Oxidation State". Journal of Chemical Education. 88 (3): 282–283. ISSN 0021-9584. doi:10.1021/ed1005213

- Phillips, John; Strozak, Victor; Wistrom, Cheryl (2000). Chemistry: Concepts and Applications. Glencoe McGraw-Hill. p. 558. ISBN 978-0-02-828210-7

- Huot, J. Y. (1989). "Electrolytic Hydrogenation and Amorphization of Pd-Zr Alloys". Journal of The Electrochemical Society. 136 (3): 630. ISSN 0013-4651. doi:10.1149/1.2096700

- Jensen, William B. (2011). "Oxidation States versus Oxidation Numbers". Journal of Chemical Education. 88 (12): 1599–1600. ISSN 0021-9584. doi:10.1021/ed2001347

- Cotton, F. Albert; Wilkinson, Geoffrey; Murillo, Carlos A.; Bochmann, Manfred (1999), Advanced Inorganic Chemistry (6th ed.), New York: Wiley-Interscience, ISBN 0-471-19957-5

- "Project Details: Towards a comprehensive definition of oxidation state". IUPAC. 2016. Retrieved December 29, 2016

- Calvert, J. G. (1990). "Glossary of atmospheric chemistry terms (Recommendations 1990)". Pure and Applied Chemistry. 62 (11): 2167–2219. ISSN 0033-4545. doi:10.1351/pac199062112167

- Frost, Arthur (1951). "Oxidation Potential-Free Energy Diagrams". Journal of the American Chemical Society. 73 (6): 2680–2682. doi:10.1021/ja01150a074

- Huizenga, John R. (1993). "Cold Fusion: The Scientific Fiasco of the Century" (2 ed.). Oxford and New York: Oxford University Press: 33, 47. ISBN 0-19-855817-1

- Karen, Pavel; McArdle, Patrick; Takats, Josef (2016). "Comprehensive definition of oxidation state (IUPAC Recommendations 2016)". Pure and Applied Chemistry. 88 (10). doi:10.1515/pac-2015-1204

- Martínez de Illarduya, Jesús M.; Villafane, Fernando (June 1994). "A Warning for Frost Diagram Users". Journal of Chemical Education. 71 (6): 480–482. doi:10.1021/ed071p480

- Greenwood, Norman N.; Earnshaw, Alan (1997). Chemistry of the Elements (2nd ed.). Butterworth-Heinemann. ISBN 0-08-037941-9

Permissions

Index

A

Acid-base Equilibrium, 129

Acid-base Reaction, 79, 122, 124, 126, 129, 131, 143, 155

Alcohol Dehydrogenase, 61

Antiferromagnetism, 24, 26, 29

Aromaticity, 99

Arrhenius Theory, 122, 127, 130-131

Associative Processes, 9

B

Basic Media, 162

Bioinorganic Chemistry, 10, 35-37

Boiling Points, 70, 77, 82-83, 98

Boltzmann Factors, 185

Bond Strength, 71, 75, 115

Brittleness, 83

Brønsted-lowry Definition, 124-125, 135

C

Chelating Effect, 15

Chemical Bonding, 20, 64, 71, 100-101, 116, 141

Chemical Hardness, 146

Cis-trans Isomerism, 5

Complex Lewis Acids, 138

Comproportionation, 176, 179-180

Coordination Complex, 1, 3, 16, 19, 27, 70

Coordination Compounds, 9, 12-13, 16, 22

Covalent Bond, 3, 18, 28, 65, 67-70, 72, 75, 96-98, 101-103, 126, 134, 141-142

Covalent Structures, 97-98

Crystal Field Stabilization Energy, 22, 111

Crystal Field Theory, 6, 20, 23, 31, 115

Cuprous Ion, 192

Cytochrome C Oxidase, 36, 52-53

D

Diamagnetism, 26

Dipolar Bond, 69

Disproportionation, 175-177, 179-180

E

Electrode Potential, 156, 173-174, 182, 188, 190

Electron Transfers, 9

Electron-deficiency, 99

Electronegativity, 68-71, 73, 76, 97-99, 112, 118, 145, 147-149, 166-167, 171-173

Electronic Density, 100

Exceptions, 11, 33, 98, 109, 166-167, 172

F

Facial-meridional Isomerism, 5

Ferrimagnetism, 24, 26, 29

Ferromagnetism, 24, 26, 29

Franklin Theory, 122, 130

Frost Diagram, 174, 177-181, 193

H

Hard-soft Trends, 144

Hemocyanine, 58

High-spin, 9, 21-23, 29-31, 33, 52

Hypervalence, 99

I

Inequivalent Atoms, 169

Intermolecular Bonding, 70

Ionic Bonding, 66, 68, 70-76, 137, 145

Ionic Compound, 17, 75-80, 83, 85-86, 89-90, 167

Iron-sulfur Proteins, 36, 55-56

K

Kapustinskii Equation, 80, 91

L

Latimer Diagram, 173-175, 178, 180

Lattice Energy, 75, 80, 83, 89-93

Lewis Acids, 126, 136-140, 143, iii 147-148, 152

Lewis Bases, 18, 126, 136-140, 145

Lewis Definition, 126, 148

Liberated Iodine, 192

Ligand Exchange, 9

Low-spin, 9, 21-22, 29-31, 33, 52

M

Magnetic Dipoles, 27

Magnetism, 33

Memory Aids, 163

Metal Displacement, 158

Metallic Bonding, 66, 70, 74, 76

Molecular Orbital Theory, 7, 20, 65, 67, 71, 99, 114-115, 117, 121

Multiple Bonds, 69, 97, 105

N

Nernst Equation, 182-185, 188-189

Neuron Activity States, 45

Nitrogenase, 37, 51, 55, 57

O

Odd-electron Molecules, 111

Optical Properties, 23

P

Paramagnetism, 24, 26-28, 30, 32, 98, 114

Photosystems, 49

Polarization Effects, 75, 92

Polydentate, 1, 10, 48

Q

Quantum Mechanical Description, 99

R

Redox Cycling, 161

Redox Processes, 154-155

Redox Reactions, 9, 86, 153, 157, 159-162, 177, 181, 190

Reducing Agents, 155-156

Reduction-oxidation Reaction, 153

Resting Potential, 44, 184

S

Signal Transducer, 44-45

Sodium Thiosulfate, 13, 192

Sodium-potassium Pumps, 43

Solvent System Definition, 126, 128

Spectroscopic Oxidation, 172

Stability Constant, 11-12

Standard Electrode Potentials, 156, 174

Stereoisomerism, 4

Structural Isomerism, 6

T

Theoretical Treatments, 90

Three-electron Bonds, 98

Transition Metal Molecules, 110

V

Valance Theory, 18

www.ingramcontent.com/pod-product-compliance
Lightning Source LLC
Chambersburg PA
CBHW082014190326

41458CB00010B/3184